T0215353

Unpacking Pedagogy

New Perspectives for
Mathematics Classrooms

A volume in
International Perspectives on Mathematics Education: Cognition, Equity, & Society
Bharath Sriraman and Lyn English, *Series Editors*

International Perspectives on Mathematics Education: Cognition, Equity, & Society

Bharath Sriraman and Lyn English, *Series Editors*

Challenging Perspectives on Mathematics Classroom Communication
  Edited by Anna Chronaki and Iben Maj Christiansen

Mathematical Representation at the Interface of Body and Culture
  Edited by Wolff-Michael Roth

Mathematics Education within the Postmodern
  Edited by Margaret Walshaw

Research Mathematics Classrooms: A Critical Examination of Methodology
  Edited by Simon Goodchild and Lyn English

Which Way Social Justice in Mathematics Education
  Edited by Leone Burton

IN DEVELOPMENT

International Perspectives on Gender and Mathematics Education
  Edited by Helen J. Forgasz, Joanne Rossi Becker, Kyunghwa Lee, and Olof Steinthorsdottir

Unpacking Pedagogy

New Perspectives for Mathematics Classrooms

Edited by

Margaret Walshaw
Massey University

INFORMATION AGE PUBLISHING, INC.
Charlotte, NC • www.infoagepub.com

Library of Congress Cataloging-in-Publication Data

Unpacking pedagogy : new perspectives for mathematics classrooms / edited by Margaret Walshaw.
    p. cm. – (International perspectives on mathematics education: cognition, equity, & society)
 Includes bibliographical references.
 ISBN 978-1-60752-427-4 (pbk.) – ISBN 978-1-60752-428-1 (hardcover) – ISBN 978-1-60752-429-8 (e-book)
1. Mathematics–Study and teaching. 2. Critical pedagogy. I. Walshaw, Margaret.
 QA11.2.U56 2010
 510.71–dc22

                                   2010000943

Printed in the United States of America

CONTENTS

SECTION 1

PSYCHOANALYTIC APPROACHES TO PEDAGOGY

SECTION 2
DISCURSIVE APPROACHES TO PEDAGOGY

SECTION 3
INTEGRATED APPROACHES TO PEDAGOGY

SERIES EDITORS' FOREWORD

This edited volume follows directly from an earlier collection, *Mathematics Education Within the Postmodern*, published in 2004 by Information Age. Given the positive response from readers of the earlier volume, this collection of articles is also grounded in postmodern theories of the social. It is focused on the specific theme of mathematics pedagogy. This volume represents a serious attempt to understand what it is that structures the pedagogical experience. In that attempt, there are two main objectives. One is a theoretical interest that involves examining the issue of the subjectivity of the teacher and exploring how intersubjective negotiations shape the production of classroom practice. A second objective is to apply these understandings to the production of mathematical knowledge and to the construction of identities in actual mathematics classrooms. To that end, the book contains substantial essays that draw on postmodern philosophies of the social to explore theory's relationship with the practice of mathematics pedagogy.

Unpacking Pedagogy takes new ideas seriously and engages readers in theory development. Groundbreaking in content, the book investigates how our thinking about classroom practice, in general, and mathematics teaching (and learning), in particular, might be transformed. As a key resource for interrogating and understanding classroom life, the book's sophisticated analyses allow readers to build new knowledge about mathematics pedagogy. In turn, that new knowledge will provide them with the tools to engage more actively in educational criticism and to play a role in educational change.

Bharath Sriraman
Lyn English

Unpacking Pedagogy: New Perspectives for Mathematics Classrooms, page vii
Copyright © 2010 by Information Age Publishing
vii

ACKNOWLEDGMENTS

Heartfelt thanks to all the contributors whose enthusiasm for new ideas has made the production of this volume so immensely uplifting and enjoyable. Wise counsel offered by a number of friends and colleagues has helped shape the book. In particular, I wish to thank the following people who have so generously given of their time as reviewers. Their incisive readings of the chapters and thoughtful comments are most appreciated.

Bill Atweh	Phil Clarkson	Peter Grootenboer
Jane Barber	Jenny Coleman	Tansy Hardy
Bill Barton	Paul Ernest	Robyn Jorgensen
Andy Begg	Jeff Evans	Mary Klein
Margaret Brown	Jayne Fleener	Heather Mendick
Seth Brown	Marg Gilling	Judy Mousley
Tania Cabral	Merrilyn Goos	Jim Neyland

Special thanks to the series editors, Bharath Sriraman and Lyn English, for their encouragement in the conceptualization of this volume and to Laurie Marshall, of Massey University, New Zealand, for providing the necessary secretarial assistance with the manuscript.

—Margaret Walshaw

INTRODUCTION

NEW PERSPECTIVES ON PEDAGOGY FOR MATHEMATICS CLASSROOMS

Margaret Walshaw

Welcome to *Unpacking Pedagogy: New Perspectives for Mathematics Classrooms.* This book represents a serious attempt to understand and explain pedagogy from beyond the standard categories of thought in mathematics education. In rethinking pedagogy, the volume emphasizes those elements of practice that are characterized not only by the regulatory practice of teaching, but also the uncertainties of practice—both inside and beyond the classroom. We interrogate mathematics teaching as a construct, situated within institutions, historical moments, as well as social, cultural, and discursive spaces. In this formulation, identities, social conditions, and political dimensions are all highly significant. In bringing the teacher's practice, in all its complexity, to center stage, *Unpacking Pedagogy* functions as an instrument for laying bare previously unthought-of aspects of practice.

Our focus on mathematics pedagogy as our object of attention is deliberately aimed at subverting stereotypical images. Portrayals within mass-

Unpacking Pedagogy: New Perspectives for Mathematics Classrooms, pages xi–xxvii
Copyright © 2010 by Information Age Publishing

mediated and ideological constructions of the roles and functions of mathematics teaching assume an essentialist character, offering a set of myths through which transmission strategies and passionate attachments to the textbook often come to the fore. In rendering this familiar story problematic, we magnify understandings that are at odds with naturalized notions of the teacher as exemplifying predetermined skills and acquiring certain dispositions. In our view, pedagogy is constantly mobile, closely linked to interactions between people, past and present, and these interactions are situated in relation to the teacher's own biography, current circumstances, investments, and commitments in teaching. In foregrounding the complexity of pedagogical practice, our critiques draw attention to the multiple dimensions and conflicting discourses surrounding the meaning of mathematics teaching that prevent us from generalizing across settings and across learners.

We are keenly aware that mathematics pedagogy is the key resource for developing students' mathematical proficiencies and their mathematical identities. But we are also aware that, as a broad discourse, mathematics pedagogy has become a "floating signifier," appropriated by various groups for different purposes. It is, for example, a gatekeeper to lifetime opportunities, signifying upward mobility and meritocracy for the successful individual student. On a wider scale, it is a major instrument of economic and social policy for achieving national objectives, particularly in a landscape in which societies confront changes in the nature of work and the structure of employment, economic recessions, and technological change. Mathematics pedagogy, in this signification, is a "political panacea." It is no surprise, then, to discover that mathematics pedagogy absorbs a large proportion of the resources of education systems—resources of people, of finance, of time, and of concrete materials. What teachers do in their mathematics classrooms is a central concern in any discussion of education.

Recent reforms throughout the world have endorsed the view that pedagogy must change if we are to counter the effects of systemic underachievement within formalized mathematics classes. We need look no further than international test data to find a sobering counterpoint to claims articulated within some sectors that all is well with mathematics teaching. The harsh reality is that many students do not succeed with mathematics. However, many teachers, searching for effectiveness and inclusiveness, struggle in their attempts to make a difference to all. On a day-to-day basis, they deal with diverse learner cohorts, and are expected to minimize the effects of the differing behavioral and epistemic responses that go hand in hand with those cohorts. Not only that: Typically, they must also contend with heavy workloads, new technologies, and new curricular policy mandates.

These problems and difficulties are nested within a much larger, complex social, cultural, and economic phenomenon. In a context in which mathematical proficiency is the cornerstone of a student's self-empower-ment, teachers and teaching have become objects of scrutiny and critique. Students' lack of proficiency is, in the eyes of policy makers, to be blamed on teachers, within the context of their infrastructure, their networks, and their institutions. Increased surveillance, set within a new audit culture, is the order of the day. Demands for evidence-based practices, scientistic pedagogical methods, and increased testing, operate within an ideological context and through normative discourses that privilege certain fundamen-talist interests, values, and practices.

In the US, for example, hard on the heels of the No Child Left Be-hind Act of 2001, funding for schools is placed squarely on a level with heightened standardized mathematics and reading performance mea-sures. Attempts have also been made in other countries to script teacher–student interactions, instructional approach, and the kind of mathematics constructed within the teaching/learning context, all of which have led to highly technicized enactments of pedagogy. Add to this the effects on teachers of marketization, and the effects on schools of competition for students, and a picture emerges of pedagogical competency as servicing the political economy of the nation state, set within the context of mobiliza-tions of powerful groups through which complex and ongoing social issues are invested and filtered. Trapped within a law of diminishing pedagogical returns, mathematics teachers struggle for social expression against hierar-chies of power and against their own marginalization.

These regimes of power—to borrow from Foucault—provide a backdrop to the research stories in this book. What it means to teach mathematics in such moments lies at the core of this book. We believe there are other ways to think about mathematics pedagogy than simply through the lens of pedagogical deficiency and (in)effectiveness. To that end, this book is a specific initiative to develop a research base for exploring mathematics teaching in new ways. As a strategic intervention, it is nested within a wider commitment to establish a postmodern presence within mathematics edu-cation, with a view toward providing creative solutions for, and new ways of looking at, interpreting and explaining phenomena. In our view, theories of action, of actors, and of agency embedded in the discipline's modernist explanations are limiting, in that they do not explain mathematics peda-gogical practice at a level that we deem desirable and productive.

The postmodern vocabulary for talking about pedagogy within this epistemic context is more relevant than ever. *Unpacking Pedagogy* takes that vocabulary seriously and engages readers in a response to context through theory development. In grappling with the questions that descend from oth-er accounts of pedagogy, we open up another "conceptual space" for talking about teaching in a way that is responsive to the complex and sometimes conflicting demands placed upon teachers, and the predictable, as well as the unexpected, contexts in which they find themselves. Working from the premise that new ideas are too important and complex to be ignored, we speak about pedagogy in ways that readers may not have imagined possible, promoting local voice and critical thinking, even as we hold critical thinking itself up for scrutiny.

We begin with the conviction that negotiating through the epistemological indeterminacy of the postmodern moment can be facilitated by looking closely into the concrete, material, and human specificities of the pedagogical experience. We do this by taking a carefully sequenced approach to the introduction of important ideas from a wide range of thinkers, and in the process, illustrate shifts in emphases from available stories of pedagogy. In principle, these emphases allow us to think afresh about research and pedagogical practice. At the same time, however, they also create a less certain space for both. However, the descriptions and explanations of pedagogy are offered in the belief that new ideas have the potential to make a substantial contribution to the formation of new forms of socio-political organization. In the developments of how the ideas might be used, we follow in the footsteps of others within the wider social sciences, in attempts to provide systematic approaches to the study of society that are able to analyze previously unthought-of and/or unexplained processes.

The book showcases both the commonalities we share amongst our thinking, as well as our differences, in terms of intellectual commitments. It represents the culmination of many combined years of work. Over a period of years, we have, both individually and collectively, disseminated our research at mathematics education national and international conferences (e.g., ICME, MERGA, MES, PME),[1] and presented and offered interpretations at discussion groups, seminars, symposia, and professional addresses. We, and a small number of others interested in what new social theories might mean for mathematics education, have worked at raising awareness of the relevance of new ideas for the discipline. Our publications have appeared in mathematics education journals (e.g., ESM, FLM, JMTE, JRME),[2] conference proceedings, and in edited collections (e.g., Brown, 2008; de Freitas & Nolan, 2008; Walls, 2009; Walshaw, 2004). All these published expositions of our intellectual work provide a background for the discussion within this volume. On rare occasions, individual authors have enjoyed the luxury of face-to-face dialogue, but more often than not, due to our geographical spread across the world, our discussions have been through email contact—always through a shared commitment to support each other's work.

The combined work of the authors in this book is centered around the development of a sound basis for their research into pedagogy. It represents a major research initiative by the group, as a whole, providing a set of theoretical, narrative, empirical, and practical applications of postmodern concepts to and around the field of mathematics pedagogy. At the heart of our analyses lies an interest in postmodernism as a "system" of ideas and as a way of understanding contemporary social and cultural phenomena. In the past, authors within mathematics education have typically focused on producing universal checklists for teaching, or prescribed solutions, or have centered their analyses on examples from "sanitized" classrooms. In the process, many crucial aspects of the pedagogical relation have remained unquestioned. In particular, discussions of classroom practice have tended to gloss over intersubjective negotiations that take place in the development of teacher identity and in classroom knowledge construction.

Accompanied by a deep concern with making the teaching of mathematics a productive activity for all students, we map out a different research-based story of teaching. In understanding the structure of teaching as inherently political, we chart teaching practice and the way in which a teaching identity evolves, tracking reflections; investigating everyday classroom activities and tools; analyzing discussions with principals, mathematics teachers, students, and educators; mapping out the effects of policy; and so forth. Our interrogations look at the lived contradictions of mathematics teaching, encompassing the struggle for self within wider meanings of and investments in the mathematics teacher. In seeking inspiration from disciplinary discourses typically beyond the frontiers of mathematics education, we trace the space of possibilities for new ideas about pedagogy in mathematics education, proposing concepts and constructs designed to investigate an increasingly complex, plural and uncertain world. Our analyses address different kinds of questions, such as:

- What are the implications of Žižek's understanding of Symbolic authority on teachers' pedagogical change during in-service initiatives?
- What does the policy characterization of mathematics as masculine mean for pedagogy?
- How does a structured enactment of the Lacanian understanding of the subject as always already rhetorically marked, influence teachers' confidence in their developing content knowledge?
- In what ways are teachers constituted as subjects of the regulatory practice of compulsory standardized testing in primary/elementary schools?
- How do classroom discourse, identity, and learning intersect to constitute the learner's fragile sense-of-self?
- In what ways do social and structural processes and material conditions interact to shape the identities of those learning to teach?
- How does the idea that "language positions people" operate within the classroom?
- What formative events of mathematics teaching and learning spark the tensions and conflict within a classroom?

- How can we rethink the notion of teacher reflection so that teachers might become comfortable with ambiguity and multiplicity?
- In what ways do digital media shape learning differently, and what pedagogic potential does this difference imply?
- How might teachers embody and enact a living pedagogy that is able to harmonize tensions among different perspectives on mathematics?
- What are the grounds for establishing a critical stance in the face of seemingly unfair access to mathematics schooling in an impoverished setting?

Our critical interrogations are at odds with the prevailing negativity surrounding mathematics teaching and learning. We believe that pedagogy's bad press represents a view that masks what is really happening in classrooms and schools. We demonstrate that point in a theoretically and empirically nuanced manner, by making visible the connections between mathematics teaching and the political and ideological arrangements in which it is nested. Revealing how the dynamics of teaching and learning intersect with forces behind the classroom, the book sensitizes us to our taken-for-granted assumptions and practices. It does this, specifically, by factoring in the ways of knowing and thinking, language, emotion, and discursive registers made available and generated within the physical, social, cultural, historical, and economic community of practice in which mathematics teaching practice is embedded.

THE IMPORTANCE OF NEW IDEAS ABOUT PEDAGOGY

Ideas about pedagogy are important. Without theoretical ideas, we would be unable to tell which aspects of pedagogical reality are critical to us and which are unimportant. Theoretical ideas allow us to understand the world of pedagogy more acutely. As a fundamental aspect of the fabric of mathematics teaching and learning, ideas about pedagogy underwrite what we understand, hope, and strive for. We go as far as suggesting that the conceptual frames we use to make sense of pedagogy have consequences for how we go about our work. For example, a focus on pedagogy through models of pedagogical deficiency stems from particular theories that guide us toward an understanding that there is something amiss about teachers themselves. In turn, that kind of understanding determines the set of questions that we might ask, and delimits the vision of the changes that might be necessary.

Recent theoretical attachments to socioculturalism (e.g, Gee, 2001; Lave, 1988) have provided us with another lens, reorienting questions about teaching from pedagogical deficiencies to positioning teaching within social practices. While there is a certain current capital in working with sociocultural ideas, as much as we would want to think to the contrary, the theory, just like any other, cannot bring everything into focus all at once. By putting a positive spin on the social and cultural aspects of reality, sociocultural theory causes us to ignore other important details about the nature of teachers' work. It blinds us from seeing, for example, the struggle for teacher's voice. Nor does it allow us to see the part that unconscious processes play in the constitution of teaching identities. That is not to say that the theory is not useful. It has, we would argue, allowed us to think afresh, to imagine things differently, and to ask new kinds of questions. The point we want to emphasize is that any theoretical lens for understanding pedagogy puts boundaries around the scope of our vision.

In this volume, seeking to cultivate an appreciation of variety and difference, we have presented a broad range of different ideas for understanding and explaining pedagogical practice. All these ideas have a close allegiance with wider epistemic shifts, and hence, highlight variable pedagogical realities embedded in social, economic, political, and historical events and processes. Irrespective of theoretical stance, across the volume, the crucial issue is rethinking pedagogy and establishing a defensible mode of inquiry that is transparent about its purpose and clear about the kinds of subjectivity and agency that obtains. In this collection, Derrida's work on deconstruction of taken-for-granted understandings; Žižek's explanation of how identities are constructed in relation to the other; Bourdieu's exposition of how everyday decisions are shaped by dispositions formed through prior events; Fairclough's insights about the way in which language produces meanings and positions people in power relations; Wilber's depiction of the evolutionary and integrated dimensions of self, culture, and nature; Foucault's understanding of how practices are produced within discourses; Lyotard's explanations of language games as fundamental to the social bond; and Gadamer's insistence on interpretation as an ongoing process are all highly influential.

Elaborating on and putting new theoretical models to use, the authors demonstrate affiliations with specific theorists, and in doing so, reconsider pedagogy through a unique lens. Rather than proposing the holy grail of mathematics pedagogy, each author's theoretical decisions have been made on the basis of the theory's explanatory power for highlighting and explaining particular aspects of the pedagogical story. As a collective, the volume does not provide a seamless analytic consensus, but what remains consistent, however, is that each analysis offers a new way of thinking. While the different analyses have few concepts in common, all rely on the underlying assumption in the usefulness of new ideas for exposing aspects of practices previously situated beyond our vision. Each is committed to approaches to pedagogy that question given understandings. Each offers new conceptual toolkits for exploring the practices and processes surrounding mathematics pedagogy. Importantly, each provides a form of critique that acknowledges its own complicity in the analysis.

In initiating and applying new ideas, each chapter purposefully avoids a critical evaluation of the arguments the chosen theorist puts forward. In our view, many of the ideas in social theory operate at a high level of abstraction and, in themselves, create difficulty in comprehension. What this book offers, specifically, is a beginning appreciation of the depth of new insights and the way these insights offer an alternative approach to thinking about and analyzing pedagogy. Chapter authors do not attempt to form liaisons with the work of other social theorists described and applied in the book. This is because the scope of the book, in large measure, aspires to the development of a culture of appreciation of contradictory voices, counter-narratives, and competing understandings. As an ensemble of voices, our stories are enacting the Foucauldian (1970) observation that ideas shift and move. Theorizing beyond the traditions of mathematics pedagogy, we are working at enlarging the scope and rewriting the terms of analysis. The point in doing this is to suggest how it might be possible to produce new knowledge about pedagogy through exploration and critical reflection on the processes and structures of mathematics education in contemporary society.

The analyses explore lived pedagogical experience, not in the sense of capturing reality and proclaiming causes, but of understanding the complex and changing processes by which subjectivities are shaped. Working at the theoretical margins of a political project in mathematics education, our stories do not seek to legislate over the constitution and nature of pedagogical reality. Indeed, we are suspicious of persuasive true, rational, and "knowable discursive renderings" (Lather, 1998, p. 495), symptomatic of our present regime of knowing and meaning. Rather, we work at illuminating the dynamics of practice—how meanings are validated, and whose investments they privilege—sensitizing others to oppressive conditions, and highlighting possibilities for where and in what ways pedagogical practice might be changed. In effect, the book exposes the contradictory realities of teachers and the complexity and complicity of their work.

Rereading the practices and processes in mathematics education through different conceptual language allows us to scrutinize the rules and practices of education. In seeking to capture the fluidity and complexity of identity constitution of teachers, the chapters, taken together, reveal how different contexts carve out their own borders, and how each represents different and competing relations of power, knowledge, dependency, commitment, and negotiation. The authors provide accounts of the tentative and shifting balance between theory and classroom experience, and the recurring tension between curriculum and the emergent personal relationships and meanings. Hence, the chapter authors do not provide victory narratives of mathematics pedagogy. Standing up against discourses premised on remediation and salvation, the stories reveal a commitment to engage in political struggle over the meaning of pedagogy itself, while simultaneously acknowledging that to speak of transformative change is to question the very meanings of empowerment.

Putting theorists to work, as we have done in this book, we attempt to map out contextualized pedagogical responses to actual classroom dilemmas, drawing on our own everyday work as educators and educational researchers. These initiatives allow us to make choices about how to speak, and write, and teach in ways that move toward the kind of arrangements in mathematics education that are more desirable for the concrete and geographical settings and material conditions in which teachers find themselves at particular historical moments. Any changes are always limited, of course, but analyses like these can help create an opening in which knowledges, roles, and relationships are questioned, where new possibilities are envisioned—an opening in which we reflect on where we are today in our understanding of pedagogy, and how we have come to be this way, and the consequences of our thoughts and actions.

What the book offers is "a praxis without guaranteed subjects or objects, oriented toward the as-yet-incompletely thinkable conditions and potential of given arrangements" (Lather, 1998, p. 488). These are the very conditions and arrangements—those that we cannot articulate at the present time, transgressing and exceeding a knowable order—that operate as enabling sites for pedagogy within mathematics education. We believe that such a praxis is ripe for development within both the intellectual conditions and the material settings of our discipline.

ORGANIZATION OF THE BOOK

The volume represents a comprehensive, but integrated, introduction to new ideas and their application to pedagogical practice across the levels, from primary/elementary school to adult education. The structure of the book is designed to assist readers to come to grips with new ideas and to see how those ideas might be used in a systematic way to explore pedagogy in local sites. The approach we take is through both exposition and application, allowing us to sketch out a rich tapestry of ideas in broad strokes. Specifically, each chapter has both a theoretical and a practical interest, and consists of two parts: (1) a focus on theory that involves examining key concepts and thinking, and (2) a practical interest that applies those same concepts to pedagogy within specific historical, cultural, and social contexts.

An *Introducing Ideas* section at the beginning of each chapter is directed toward cognitive mapping to new selected conceptual language for understanding pedagogy. It provides an overview of the thinking and key concepts of the theorist(s) that have shaped the discussion in the chapter. The *Applying Ideas* section of each chapter explores those theoretical insights, in relation to actual mathematics teaching experiences, settings, or processes. The *Introducing Ideas* and *Applying Ideas* sections are separated merely for the purpose of enabling easier access to new ideas. In our view, theory and practice are intimately connected—all pedagogical practice is informed by, and, in turn, informs theory.

The theoretical focus within the chapters has served to organize the order of the chapters, and has led to the establishment of three main sections. Chapters are arranged according to psychoanalytic approaches, discursive approaches, and integrated approaches.

Psychoanalytic Approaches

In Chapter 1, Una Hanley explores how teachers make sense of and enact curriculum reform. The location is England, and the context is one of constant curriculum change, of a proliferation of official documentation, of widespread testing, and of assessment measures that tend to foster transmission approaches. In an effort to move teachers from technicist models of teaching, a research project was developed involving sixteen teachers from six schools. In this project, subjectivity is a central concept, and Una uses psychoanalytic resources to explore its constitution in a group of teachers. The researchers shared knowledge about, and approaches to, working with new materials. An apprenticeship model was sought as the teachers represented varying ages and varying years of teaching experience. Introducing and drawing on Žižek's understanding of Symbolic authority, Una reveals how, for the teachers in this project, subjectivity is at the crossroads, highlighting the force of power in legitimating investments within certain psychic realities. She shows that while teachers attempted to put into practice what they learned through the research project, for some, teaching practice was merely a performance—always with one eye on what the researcher wanted, and the other eye focused squarely on the "other." What is learned and practiced in research initiatives is sometimes never fully cashed in as an educational capital within the classroom. The exploration sketches out the challenges in fully realizing the objects of our research endeavors and, specifically, the difficulties involved in negotiating pedagogical change when teachers are invested in other ways of thinking and acting.

In Chapter 2, Tamara Bibby helps us come to terms with concepts within psychoanalytic theory and shows how they might be used in scholarly work on the pedagogical relation. She draws on the concepts of the oedipal family and the Oedipus complex to unpack relationships to mathematics, particularly as they are constituted in primary schools. Post-Freudian psychoanalytic theories of authority provide her with the conceptual tools to investigate the way in which mathematics, with an emphasis on rules, speed, and correct answers, is characterized as masculine in traditional school mathematics pedagogy. Taking care not to essentialize gender, Tamara unpacks the ideational fiction of binary characterization, and proffers, instead, masculinity and femininity, boy and girl, as "elements within gender." Tamara draws on research data from UK projects to unpack some of the potential consequences of differentiating mathematics as an unemotional, authoritative, rational, systematic, and logical set of values and practices, away from so-called feminine qualities, such as warmth, emotional attunement, and creativity. Specifically, she explores the tensions that result from fictions that allow for the deployment of masculinity in the discursive construction of mathematics and investigates the consequences for teachers and students living with the effects of these splits in policy and practice. In seeking a way in which to reconceptualize mathematics differently from masculine overtones, Tamara suggests framing the experiences of learners through the Oedipus complex, as a way of dealing with the production of students' subjectivities.

In Chapter 3, Tony Cotton's focus is on the development of pre-service teachers' subject knowledge for the purpose of enhancing teaching practice. Under interrogation is Shulman's notion of pedagogical content knowledge. Tony investigates an initiative in the UK that took the form of a series of seminars, involving 12 pre-service teachers who identified as "lacking in confidence" in content knowledge. Underpinning the discussion is a concern with the question of identity and the difficulty that pre-service teachers have, new to the profession, in tracing out for themselves an identity as a teacher. Implicitly arguing for the unconscious construction of subjectivity in preference over the stable empirical self, Tony works with the material of teachers' narratives, and paying attention to language, shows how a psychoanalytic logic produces new identities for these teachers. Using ideas drawn from Lacan and others—specifically, the idea that identity is always relational—Tony explores identity formation at work with the group in the research initiative. He probes the self–other relation through intersubjective negotiations that are made possible in the research. Through a structured process, involving a self-conscious critique of their own constructions of themselves as teachers, their views of mathematics, and their learning of mathematics, the pre-service teachers come to a realization of how their understandings surfaced and impacted their teaching practice. The process and its findings lead to an alternative understanding of pedagogical content knowledge.

In Chapter 4, Fiona Walls unpacks perceptions of the "good" teacher, setting her exploration within a climate of compulsory standardized testing. Within that context, she draws links between the testing regime the monitoring of the school and teachers' practices. Using ideas drawn from psychosocial theory, she underwrites the discussion with the Foucauldian understanding that power seeps through institutional settings, governing and regulating practice in schools, and normalizing what teachers do. She also draws on the notion of teacher identity as a process embedded in discourse, and uses Lacan's idea of the self as formed with a view towards what the other wants. Using data from research with nine teachers of Grades 5 and 7 within an Australian school, Fiona reconfigures the relationship involving identity, experience, desire, and official discourse. Mapping out the way in which teachers negotiate their way through contesting perceptions of teaching, Fiona shows how systemic forces are lived as individual dilemmas in which teachers embody practices that they had wanted to change. With a focus on why and how teachers structure their teaching identities in the way they do, she unpacks the force of feelings of the teachers' "self" towards their imagined "other," highlighting the way in which teachers speak of their highly compromised (and limiting) practice within mathematics classrooms.

Discursive Approaches

In Chapter 5, Diana Stentoft and Paola Valero investigate the fragility of mathematical learning. Their discussion expresses a poststructuralist imagination that takes seriously the notion that language constitutes social reality, rather than reflecting an already given reality. In developing an understanding of the "noise," symptomatic of everyday classrooms, their view is towards challenging interpretations of the practices within what are typically characterized as "pure mathematics classrooms." Their theoretical approach uses precepts that are, in tenor, at odds with the presuppositions that ground the rational autonomous learner. In the discussion, undermining traditional approaches to learning, Diana and Paola draw attention to the interrelatedness, as well as the fragility of, classroom discourse, identity, and learning. They argue that these three elements together constitute the landscape within which students' sense-of-self as learners is formed. In their discursive analysis, they case-study mathematics classroom interactions at a Danish teacher training college. Underlying the analysis is an intent to avoid mere descriptions of classroom life, and, rather, unpack how students and teachers are involved with constructing multiple identities over the course of a mathematics lesson. The intent is also to make clear how learning mathematics and constructing mathematical knowledge in the classroom is inextricably caught up in the discursive practices of the classroom.

In Chapter 6, the interest is in finding out how pre-service teachers come to teach and learn in the context of their mathematics practice in schools. The spotlight is on the ways in which social and structural processes and material conditions interact in the shaping of those learning to teach. Introducing and applying a number of concepts drawn from Foucault's theoretical toolkit, I analyze pre-service elementary teachers' reflections of their practicum experience. These were undergraduate students working towards a bachelor's degree in education within a New Zealand university. I use the reflections as a means of understanding the local, systemic, and flexible conditions of identity construction, and the way an identity as a teacher is produced and reproduced through social interaction, daily negotiations, and within contexts that are already overburdened with the meanings of others. In revealing the way in which particular practices are normalized and monitored within the practicum experience, the political and strategic nature of teaching practice comes to the fore. Learning to teach, it is found, is an activity set within a barely visible set of highly coercive practices that target thinking, speech, and actions, focused on producing particular constructions of identity. The analysis sensitizes us to the impact of regulatory practices on pre-service teachers' understanding of themselves as teachers and on their constructions of what it means to teach effectively in the mathematics classroom. Specifically, it draws attention to the importance of relationships and forms of reciprocity and obligation inherent within the pre-service/supervisor relation.

In Chapter 7, Elizabeth de Freitas uses critical discourse analysis to study the classroom discourse and interaction patterns of two secondary school mathematics teachers of senior classes in Canada. She employs Fairclough's understanding that language not only produces meaning, but also positions speakers in specific relations of power. The purpose is to understand the way in which teachers' subjectivity is constituted and enacted, in brief and often spontaneous and contradictory speech acts. The task demands thinking about text and context in classroom interaction as intersecting, rather than separated. In the analysis, Elizabeth shows how one teacher, Mark, repeatedly used metaphors that signified an antagonistic relationship between students and texts, and embedded many references to sports throughout his lessons. She demonstrates how the other teacher, Roy, continuously made reference to the difficulty of learning calculus, choosing to exclude discourse from other texts that spoke calculus into existence in other ways. Both analyses highlight what teachers choose to say and the way in which they say it, and the power relations that descend from those linguistic decisions. In particular, the analyses provide counter-narratives about classroom discourse, pointing to the regulatory power of teacher discourse in providing access to mathematics by shedding light on those students who are included within and those who are positioned outside of the text. Importantly, through the fine-grained reading that unpacks hidden relationships and regulatory practices operating within the classroom, Elizabeth demonstrates the way in which the discursive practices of the two teachers contribute to the kind of thinking that is possible within the classroom.

In Chapter 8, Kathleen Nolan draws on Bourdieu's conceptual framework to explore her efforts as a mathematics educator at developing an inquiry-based classroom in an undergraduate teacher education program in Canada for middle years teaching. Taking a self-study approach, principled upon a desire to improve her own instructional practice, Kathleen shows how her inquiry teaching approaches, that required a tolerance of ambiguity and uncertainty from students, met with resistance, challenge, and dissatisfaction from students when it ran up against their entrenched understandings of mathematics teaching. Constructions of pedagogy made available through their encounter with teaching as former students in mathematics classrooms compete for the individual pre-service teachers' attention, and both summon and dismiss new proposals for pedagogy. Kathleen analyzes the tensions between thought and action, knowledge and experience, and the technical and existential enacted in the pedagogical encounter. She does this by paying attention to both her students' and her own *habitus,* and the *cultural capital* that each party brings to the *field* of mathematics teaching. Bourdieu's theory, woven into the self-study, provides the tools to lay bare contradictory visions and conflicting interpretations, as well as the negotiation they both demand. The story of pedagogy presented in the chapter provides an honest account of the lived tensions within the shaping of pedagogical relationships and the dilemmas in trying to establish one's authority in a context fuelled with contestation. In pointing to the hidden work demanded of conflict resolution, the chapter offers a suggestion as to why reforms in teacher education do not always enjoy an enduring effect.

Integrated Approaches

In Chapter 9, Moshe Renert and Brent Davis introduce evolutionary frameworks of complexity science to reformulate the subject matter of mathematics. They highlight Wilber's model of evolutionary dimensions that integrate self, culture, and nature, through which a plurality of perspectives may be entertained. In a creative spirit of critique, by tracking how understandings of mathematics have evolved over time, Moshe and Brent show how the alternative conceptions lead to different views about the kinds of tools that mathematics uses. Each makes it possible for certain understandings to be entertained and legitimated. Unlike proponents of a number of current perspectives, Moshe and Brent do not suggest outright dismissal of specific perspectives. Rather, they suggest that each of these views has something important to tell us about the shape and character of mathematics, even as each has its own limitations. Their proposal is towards an evolutionary perspective, one that is inclusive of the contributions of traditional, modernist, and postmodern perspectives. Their integral perspective values the enacted, creative and dynamic dimensions of mathematics, and is focused on the health and harmony of the entire system. Moshe and Brent apply these ideas to their work with eleven experienced middle school teachers. Their work demonstrates how teachers are crucial participants in the creation of mathematical possibilities. They suggest that teachers might engage students more meaningfully with mathematics by elaborating the specific, by using active language, and by allowing them to engage with multiplicity and plurality in discourse, meaning-making, and interpretation.

In Chapter 10, David Stinson and Ginny Power explore the concept of teacher reflection, as it applies to a group of practicing teachers enrolled in a graduate-level course. Putting to use, as well as troubling, Dewey's concept of reflective thinking, and drawing on Foucault's notion of discourse, Derrida's concept of deconstruction, and the understanding of *rhizome* as offered by Deleuze and Guattari, they redefine the notion of the reflective practitioner in order to capture the dynamic between subjectivity and educational practices. Their central theme pivots around teachers' changed thinking as a result of engagement with different texts. In their discussion, David and Ginny make a case for introducing mathematics teachers to postmodern thought, particularly in an era that is heavily focused on teacher surveillance and regulation. Supporting their position with extracts taken from tertiary students' journals and reflected essays, they sketch out the ways in which different understandings about mathematics teaching, which neither stretch plausibility nor break with reality, emerge through the teachers' appropriation of the ideas within the texts. They show how exposure to and engagement with postmodern ideas leads to significant changes in the teachers' thinking about pedagogical practice. Interweaving comments written by the students with concepts borrowed from postmodern philosophers and theorists, they illustrate how the teachers began to understand that teachers and students might, indeed, be described differently in the postmodern.

In Chapter 11, Nigel Calder and Tony Brown explore students' engagement with digital technologies and investigate the ways in which understanding emerges when mathematical phenomena are engaged through such media. Their focus is on the pedagogical potential, within primary/elementary classrooms, of Information and Communication Technologies (ICT), and specifically, the potential of the spreadsheet. Adopting Gadamer's hermeneutic perspective to meaning-making and applying it to learning, Nigel and Tony argue that mathematical learning comprises a process of interpretation, where understanding and "concepts" might be seen as states caught in ongoing formation, rather than as once-and-for-all fixed realities. What comes to the fore is the merging of concrete practical tasks with hidden cognitive activity. Understandings of mathematical phenomena, and of who we are, evolve through cyclical engagements with the phenomena and through constantly drawing forward prior experiences and understandings. The pedagogical medium, the mathematical task, the pre-conceptions of the learners, and the associated dialogue evoked are interdependent and co-formative. Nigel and Tony apply these ideas to data drawn from a study involving 10-year-old New Zealand students to illustrate how students' learning was fashioned by the particular affordances of the pedagogical media employed. What emerges in the analysis is the pedagogical potential of digital media, and the way in which the spreadsheet, in particular, is able to accentuate students' visualization and exploration to enhance students' mathematical proficiency and mathematical identity.

In Chapter 12, Dalene Swanson explores the discourse of mathematics pedagogy, sketching out models of democracy, freedom, and development, as they apply to teaching and learning within the South African context. Drawing on understandings developed from narrative theory, she offers alternative meanings to these concepts, based on the authenticity of being a researcher in a Grade 7 mathematics classroom and her conversations with the principal of the school. Dalene's central argument is that humanitarian concepts that we typically hold to be true for all are, in fact, construed differently by others, and for very different purposes. For her, developing empowering approaches for learners who are polarized and disempowered by their socioeconomic status is a key concern. The analysis highlights important fundamental issues about social justice and about teaching and learning, specifically those issues to do with choosing or not choosing to teach mathematics, and, behind that issue, who has access to mathematical knowledge. Within a context in which she finds certain aspects of experiences morally unconscionable, Dalene draws attention to the way in which understandings of what is "good" for an impoverished community are constructed from ideologies that are not necessarily the kinds that she, and, presumably, many others, would endorse. She is not offering a panacea, but writes of the humbling experience, and of an ethical responsibility, in trying to see from others' point of view and of attempting to create a moment for herself within which widely disparate views might coexist more easily. Dalene's project raises questions about action and change, and about interactions that cannot be separated from axes of social and material deprivation that operate to define learners. She suggests that "empowering" approaches to the teaching of mathematics should be used cautiously.

NEW PEDAGOGICAL TERRITORY

The ideas in this collection represent new pedagogical territory. They are addressed to teachers, teacher educators, and researchers, as well as to students in education, in general, and in mathematics education, in particular. The book is intended to be read iteratively, shaping and reshaping understanding, in response to readers' own continuing questions and pursuit of knowledge about what drives classroom practice and with what effects. At the end of each chapter we offer a short list of references designed as a starting point for further reading. In some instances, these represent references to the theorists' original work, and, in other cases, the references mark out work that has synergies with the ideas introduced within the chapter. We hope that readers will follow up with these sources for the purpose of exercising the imagination, for extending knowledge, and, above all, for rethinking everyday pedagogical practice in mathematics classrooms.

NOTES

1. ICME: International Commission for Mathematics Education
   MERGA: Mathematics Education Research Group of Australasia
   MES: Mathematics Education and Society
   PME: International Group for the Psychology of Mathematics Education
2. ESM: *Educational Studies in Mathematics*
   FLM: *For the Learning of Mathematics*
   JMTE: *Journal of Mathematics Teacher Education*
   JRME: *Journal for Research in Mathematics Education*

REFERENCES

Brown, T. (Ed.), (2008). *The psychology of mathematics education: A psychoanalytic displacement.* Rotterdam: Sense Publishers.

de Freitas, E., &. Nolan, K. (Eds.) (2008). *Opening the research text: Critical insights and in(ter)ventions into mathematics education.* Dordrecht: Springer.

Foucault, M. (1970). *The order of things: An archaeology of the human sciences.* New York: Vintage Books.

Gee, J. P. (2001). Identity as an analytic lens for research in education. *Review of Research in Education, 25*, 99–125.

Lather, P. (1998). Critical pedagogy and its complicities: A praxis of stuck places. *Educational Theory, 48*(4), 487–497.

Lave, J. (1988). *Cognition in practice: Mind, mathematics and culture in everyday life.* Cambridge: Cambridge University Press.

Walls, F. (2009). *Mathematical subjects: Children talk about their mathematics lives.* Dordrecht: Springer.

Walshaw, M. (Ed.) (2004). *Mathematics education within the postmodern.* Greenwich: Information Age.

SECTION 1

PSYCHOANALYTIC APPROACHES TO PEDAGOGY

CHAPTER 1

TEACHERS AND CURRICULUM CHANGE

Working to Get it Right

Una Hanley

ABSTRACT

This chapter is about looking at teachers' experience of curriculum development through a psychoanalytic lens. I begin the chapter by setting the scene in England, where teachers are subject to a constant stream of initiatives. In a theoretical interlude, I describe the theoretical "lens" in more detail, questioning the technical–rationalist frameworks customarily employed to consider innovation. The concerns of mathematics educators are set out following this showing, illustrating the shift into paradigms outside of a technical–rationalist model, and finally, I read a set of data using the psychoanalytic framework as a guide, in order to re-think the experience of teachers undergoing change.

As education in England has become focused increasingly on measurement and centralized control, the interests and concerns of individual teachers have become marginalized by the discourses of achievement and economic viability. There is an emphasis on certainty and a rationality of a kind that

Unpacking Pedagogy: New Perspectives for Mathematics Classrooms, pages 3–19
3

regards learning as linear and easy to insert into the developmental pathways of pupils. Documentation associated with curriculum advice suggests that these pathways are predictable, incremental, and measurable. While the performance of pupils remains in the spotlight of curriculum intervention, perspectives relating to teacher development are subject to similar influences, whereby ongoing performance is to be managed and developed along prescribed routes. In England, there is a concerted move to create a framework of competences with clear descriptors of progressive levels of ability and ways to improve that are to be applied to practicing teachers. Much of this improvement lies in the development of "meta-cognitive skills," which are presumed to underpin improvements to practice, that is, an assumption that teachers respond to the teaching task primarily by "thinking it through," cognitively. In this way, performance-oriented routes to professional development tend to bypass the immediate concerns of individuals, and success or failure tends to be compartmentalized into certain categories in which "remedies" might be applied and, potentially, notable improvements made.

All disciplines in education can be thought of as tied into sets of binaries in which outside intervention is frequently tilted against the preferences of the individual. Mathematics education is no exception. The language in which these initiatives are embedded walks an unsteady line between an agenda that nominally supports increased teacher autonomy and one which overtly privileges standards and outputs with ever-increasing levels of accountability. However, teachers continue to value the former, although the spaces where this can be expressed have become scarcer. For many newly qualified teachers (NQTs), keen to be viewed as competent, a perceived route to being a successful teacher lies in a close identification with the governmental advice; in such an approach, personal attributes may add a surface sparkle to prescribed approaches.

Continuing government intervention into teaching practices and curriculum "produces" teachers as practitioners who are apparently fated to perform in deficit. Although the newly devised curriculum model is "applied," the practices of teachers appear to be conflated with curriculum recommendations when there is a failure to increase levels of achievement in line with predictions. There are a number of proffered explanations for "teacher failure." Fullan (2003) writes about the perceived muted enthusiasm of teachers toward innovation and the associated assessment instruments developed to record student performance. In relation to this dilemma, Fullan writes, "as scores rose, the morale of teachers and principals, if anything declined" (p. 3). He goes on to say, "Engaged students, energetic and committed teachers, improvements in problem-solving and thinking skills, greater emotional intelligence, and, generally, teaching and learning for deeper understanding cannot be orchestrated from the center" (p. 3). In this extended list, there can be discerned a shift from the more familiar technical rationalist agenda associated with curriculum innovation, toward rather more nebulous and altruistic features that carry personal appeal. The list here works to fix an implicitly unfinished list of qualities, which, amongst my colleagues, might be regarded as the "fairy dust" of teaching, that is, something inherently "more than" documents can convey, which relies on an emotional response, rather than a response more closely associated with methods. Overall, there is an acknowledgement that teachers' engagement with their task cannot be neatly boxed into incremental stages toward mastery of the latest innovation "orchestrated from the center." Not everything can be managed from "the outside–in."

I begin here with a rather playful/serious move to align curriculum reform with what might be described as the collective fantasies of curriculum reformers. For those of us who are both observers and participants in curriculum reform, it appears that the resolution to all the perceived difficulties involved in improving things, now specifically related to raising measurable achievement, appears to be not quite graspable in the moment. In England, the Department for Children, Families and Schools (DCFS, and its earlier manifestations, e.g., DfEE) has been responsible for over 1,600 documents issued since the year 2000. While these documents are not confined to curriculum recommendations, they offer an indication that there is a concerted search for a more effective education system, even if applied in small, incremental stages, in the main (Ball, 2008). Yet the continuing proliferation of documentation also implies that solutions to difficulties appear to remain as elusive as ever. Ball (2008) goes on to say that, while the relationship between policy and practice is complex—that is, "Policies are contested, interpreted and enacted . . . and meanings of policy makers do not always translate directly and obviously into institutional practices"—it is important not to "over-estimate the logical rationality of policy. Policy strategies, acts, guidelines and initiatives are often messy, contradictory, confused and unclear" (p. 7). Logical rationality, then, has an untidy underbelly, which is shaped discursively into something apparently manageable by the determining orchestrations "at the center," but much more difficult to realize *in situ* (Coffield et al., 2007).

In describing the tensions that exist for teachers between the "economies of performance" and "ecologies of practice" (p. 131), Stronach et al. (2002) draw attention to the notional boundary between the outside–in and inside–out dynamic. Teachers understand that students have varying inclinations, abilities, and commitment to learning, and they see these difficulties smoothed out by a curriculum that assumes a readiness amongst pupils to learn, in both cognitive and affective terms. Although it is recognized that there will be some variation, there is an implicit assumption in the available curriculum advice that students will learn along the pathways prescribed and at a pace determined by the curriculum content. However, the relationship between "inside" and "outside" is not so easily drawn. Psychoanalysis fudges the boundary between self and context—society and its requirements—so that neither is a distinct entity in relation to the other. Permeating this contested area for the individual, is the Imaginary. A theoretical feint is called for here.

INTRODUCING IDEAS

Psychoanalysis and education are perhaps uneasy bedfellows, but, nonetheless, are on speaking terms, and these terms have become more credible among educationalists, as there is an increasing recognition that technical–rational perspectives can offer only a limited picture of teachers and teaching. There are a number of perspectives employed in psychoanalysis, and I have found Žižek's take on Lacan particularly helpful, although I have borrowed from others who have worked to mediate difficult ideas, including Britzman (1998) and Easthope (1999).

In Lacanian psychoanalysis, the social context that we all inhabit is conceived of as discursive (Žižek, 1989), and each of us is constructed through and in language. Lacan (1977) specifies three domains in relation to this: the Imaginary, the Symbolic, and the Real. The Real is a complex domain, but is associated with the experiences of the individual that fall outside language. As infants, for example, we inhabit a world of being that is the domain of the Real. However, while leaving the Real order of being and entering the Symbolic order of language is both essential and unavoidable, there are penalties. Being is subjugated into meaning by the signifiers allotted to describe us, bringing about a sense of loss. The individual experiences "lack in being" (Lacan, 1977). This sense of "lack in being" fuels our desires and the fantasies we form to satisfy them.

The Symbolic order is the site of culture, replete with rules and meanings historically located that form a network of symbolic relations, and are described by Žižek (1989) as the "Other" (p. 113). The individual enters the Symbolic order attached to descriptions that give her a place in the discursive network that orientates society and from which she borrows a sense of identity. We are all affiliated to families, for example, around which certain expectations are made in relation to behavior or prospects.

However, the workings of the Symbolic are mediated by the Imaginary, which supports the individual in maintaining a sense of prowess and unity. The ego ideal—one of two aspects of the ego—provides the mechanism by which this is brought about, confirming our identity in the best possible light. The Imaginary allows us to see ourselves as we would like to be seen, as the complete individual who speaks coherently and with authority, unaware that we are the subject of many fantasies, including the fantasy that we are rational beings outside the purview of the Symbolic order.

My ego appears to be the same in space, permanent across time, and unified in substance, though all of these misrecognise how I come about as an effect, thinking I am really there, despite different spaces, my own actual dispersal into various selves, being split between conscious and unconscious (Easthope, 1999). Thus, the Imaginary works to bring together one's view of one's self, together with the view that we imagine others have of us, that is, the view of the Other.

Just as the Symbolic imposes laws and rules that give shape to life, the ego marks a conciliation of a kind between the desires and fantasies of the individual, of which he or she may be unconscious, and the rules of the Other. Lacan (1977) postulated that there was a real world but that this could only be apprehended through the laws of the Other, which come to constitute the norm, and the desires and fantasies of the individual, which, like cats in a bag, create internal conflict. The Imaginary thus becomes a site where the world of internal conflict meets the world of external discord, but where difficulties attendant to this become "smoothed over," only to erupt elsewhere. On the one hand, the eruption "elsewhere" of societal discord is only too visible, particularly to teachers, and here it needs to be noted that the Other is no more coherent and unified than the individual. On the other hand, the Imaginary actively creates the delusion of rationality within individuals, but visible personal preferences and proclivities are, nonetheless, influenced by what has been contained by the unconscious. What the ego does not wish to contemplate becomes repressed in the unconscious, re-emerging as slips of the tongue, for example, and "baffling and bungled actions" (Britzman, 1998, p. 7). This is an uncomfortable idea in an educational arena that favors the view that individuals gain mastery over their thoughts and actions.

Britzman (1998) points to the variance between the epistemology associated with education, which presumes to intervene rationally in the activities of learners, and an epistemology associated with psychoanalysis, which posits education not as an "engineered development" (p. 6), but as an interference into the activities of the actors in an educational context:

> one that insists that the inside of actors is as complicated as the outside . . . [N]ot only does the world impinge cruelly upon the subject, and not only does the subject's inner world constitute the be-all of understanding and misunderstanding: the subject lives both dilemmas in ways that cannot be predicted, authorised by another, or even deliberately planned and separated. (p. 6)

Each individual has a personal history of these dilemmas. However, a fantasy perhaps shared by the many stakeholders in curriculum development (central government, local authorities, senior management teams in school) is of (1) a classroom where "an uncommon sociality can be invented in a common place" (Britzman, 1998, p. 50), where identification with a goal toward raising achievement suppresses the effects of individual difference, and of (2) teachers who will "buy into" this fantasy.

The Objects of Fantasy

What do teachers fantasize about, in relation to their practice? It is reasonable to assume that they share the fantasies of other stakeholders, at least in relation to raising achievement? Novice teachers perhaps fantasize about a classroom in which differences can be temporarily sutured together and a space created where there is a potential for learning. Here, a little more theory is probably appropriate.

When the subject enters the Symbolic, her subjectivity is rather intangible as she tries to work out her place in the order of things, to gain a foothold, as it were, in terms of identification. However, what the Other wants is ambiguous and contradictory, but nonetheless, the individual struggles to find the parameters of the desire of the Other in order to shape herself to "fit." But there is an inevitable gap between what the Other appears to want and the individual's capacity to meet this demand. This creates a sense of "lack in being," which the subject desires to satisfy.

"Desire is an unconscious search for a lost object, lost not because it is in front of desire waiting to be re-found but because it is already behind desire and producing it in the first place" (Easthope, 1999, p. 97). Therefore, a principle of desire in a psychoanalytic context is that it can never be satisfied. Desire cannot be satisfied, but the individual finds a "stand-in" for that which is lacking, the *objet petit a*. The Imaginary shields the self from conflicts immanent to the subject's fantasies and daydreams, and provides the site where the subject settles upon the *objet petit a* that will satisfy for the time being. After a while, the subject realizes that the *objet petit a* does not "do the trick," and moves on to something else. In private life, this sense of "moving on" is very visible in the field of human relationships, where the object of desire gradually appears to lose its sought-after status. In the classroom, similar mechanisms are at work, but in a narrower framing. Here, the teacher with individual desires for an effective classroom finds a readily available stand-in in the newly proffered curriculum. However, working out what to do is far from straightforward.

Žižek describes the way in which the individual is called upon to respond to what the Other wants. Entry into language as an infant, where the individual meets the discursive frameworks of the Symbolic and is designated a place therein, is repeated as the individual enters additional discursive frameworks that carry a mandate for certain forms of behavior. Entry into a profession marks one of these occasions. Having been "called into this being," however, there is a level of bewilderment. Uncertain what to do, the individual asks "Ché vuoi?": "Why am I what I'm supposed to be? Why have I this mandate? Why am I...[a teacher, master, king]?" (Žižek, 1989, p. 113).

The individual desires to fulfill the mandate allotted to her, yet a gap persists between the subject and the space she is directed to fill. Meanings are unclear and there is always more to contend with than was initially apparent. Behind one set of meanings, there are others to be encountered. In education, the master signifier *teacher*, "calls" the subject to be the individual created in lists of competences and pedagogical styles for which a response is required, but for which there will be many variations. These difficulties are at their most visible in practice situations in which the newly qualified teacher, for example, knows that something needs to be done better than the way it currently manifests in action.

More experienced colleagues may be more attuned to "living out" their fantasies and have belief in their own rationale for action, although we all have trust in our sense of self as derived from the Other, the ego-ideal. For example, we may all have been appraised by mentoring teachers who appear to embody the very practices which they criticize in their students. Responding to a Symbolic mandate requires activity driven by conscious intention, however unclear the directions. As Britzman (1998) suggests, what the ego cannot contemplate is repressed into a domain of internal conflict, causing an additional barrier between the "call" and an individual's capacity to fulfill it. Failure to perform up to expectations may well be a feature that we all, including experienced teachers, repress. While seeing this in the activity of other people is rather too easy, repression prevents us from seeing how we act in accord within our own proclivities in our own performance. In Lacanian psychoanalysis, interpellation, the call to fulfill a Symbolic mandate, invariably fails.

With this theoretical backdrop in mind, I will be turning to a small-scale project that sought to introduce a new mathematics curriculum into six secondary schools in Greater Manchester, UK.

APPLYING IDEAS

It is now time to turn to the project in which this chapter is rooted. I have entered this project in the past as a researcher, wearing a rationalist's hat, and in that guise, I could make the assumption that curriculum intervention necessarily meant progress and that teachers could ease their diverse students along a trajectory toward improved attainment if they followed the advice correctly. On one level, given much of the enthusiasm for the new approaches that many of the teachers displayed, this is a difficult notion to relinquish. However, there were a number of attendant issues, which form the basis of the discussion in this chapter.

The mathematics curriculum in England for all age groups has recently been subjected to significant changes, but prior to this, curriculum innovation was relatively small-scale and framed around broadly constructivist approaches (Brown et al., 2007). However, the arrival of the National Curriculum in the UK in 1989 came with a detailed description of the content to be taught and the order in which it might be taught. Successive documentation, but more significantly, a tightening of assessment instruments, while recording a rise in levels of achievement, ultimately appeared to foster transmission approaches to teaching, particularly at points where teachers prepared their students for national examinations at Key Stage 2 (student age 11 years), Key Stage 3, (student age 14 years) and Key Stage 4 (student age 16 years), and where results are published nationally, school by school. Some of the initial benefits associated with these approaches have "plateaued out."

From 2004–2007, at the Manchester Metropolitan University (MMU), a small project was established that attempted to do something else. Based on in-depth discussion around a small number of problems, the approaches developed there paved the way for an extended project based on Realistic Mathematics Education (RME) and the English language version originally published in the US, Mathematics in Context (MiC).

The data presented here were derived from a government-sponsored Economic and Social Research Council (ESRC) bid, which piggy-backed on the much larger Gatsby-financed project. RME pedagogy follows a broadly socio-constructivist pedagogy, which nonetheless privileges the individual and the private realm. Mathematics in Context (MiC) presents mathematical ideas in meaningful contexts along a trajectory that has been carefully constructed toward "mathematising" amongst pupils, in which problem solving and "level raising" (generality, certainty, exactness, and brevity) are significantly privileged in an environment that fosters hypothesis and discussion. The authors describe this as a process of guided re-invention.

In RME, a trajectory is a developmental pathway that follows both a horizontal and a vertical dimension as pupils fluctuate between an understanding grounded in a specific and realizable context toward developing generalizations (van den Heuval-Panhuizen, 2000). While the writers of RME understood that the learning trajectories of pupils would be individual, there is a strong sense of a developmental pathway that carries most pupils. We felt that rather less attention was paid to the trajectories of teachers in their understanding of the new pedagogy and associated practices, and this provided the basis for the ESRC project.

The training for teachers associated with the project was implicitly embedded in a community of practice approach (e.g., Wenger et al., 2002), in which it was anticipated that experienced colleagues would support the less experienced in their endeavors to work with the new materials on five nominated study days in the academic year. Such activity is presumed to draw the less experienced toward "expertise" in a process which requires the teacher to identify strongly with both the underlying principles and the classroom materials in which these principles are embedded. The teachers who joined the project in the academic year 2005/06—the duration of the ESRC project—were from six schools representative of a range of schools in the city. The RME approach afforded an opportunity for participating teachers to consider mathematics pedagogy at a level not anticipated in government-sponsored documents. Altogether, there were 16 teachers situated in 6 schools throughout the city. Of these teachers, two were newly qualified and three had been recently qualified.

While introducing new practices is far from straightforward, the process was eased for some by association with the Centre for Mathematics Education at MMU and earlier work on approaches to classroom practice. Others did not have this experience. The ages and statuses of the project teachers varied, but each showed a willingness to interrogate his or her existing practices or an openness to extend their field of practice.

Over the period of the Gatsby project, teachers undertook to trial MiC materials in their classrooms. University tutors leading the project supplied support and made lesson observations, providing feedback. On the training days, participating teachers were offered the opportunity to review their experiences and share them with others. Through this process, they gained an understanding of the principles from which the materials were developed. The aim of the research from the ESRC-sponsored project was to focus on the experience of teachers. To do this, we supported a collaborative action research model with the participants. In the event, pressures on time meant that this process was significantly scaffolded by the research team, in the sense that entries were augmented by interviews in which participants were encouraged to reflect on their experience in greater detail than the diaries allowed for. They were asked to consider the new practices with which they were involved and to note where they felt some resonance between these and their existing beliefs about practice, and where they experienced "jars," and reflect more generally on their developing perspectives and practices in relation to MiC. These sessions were audio-taped and transcribed. Visits to school were made every six weeks, and there were additional interviews on training days at the university. In addition, our data collection was augmented by half-hour "slots" on each training day.

While the methodology for the Gatsby project was implicitly realist, the ESRC project adopted a broadly hermeneutic methodology, at least in relation to the required report.

Identifying with New Practices

The process of curriculum innovation disturbs the sense that individuals have of themselves in relation to practice. Whereas they may have felt comfortable in terms of their existing practices, in psychoanalytic terms, the symbolic mandate presented by the curricular materials makes a new set of demands that the individual is called upon to fulfill. For the first few weeks, the teachers were very concerned about delivering workable lessons as they sought to make sense of the materials, and many soon began to customize their approaches, that is, align them more closely with existing practices with which they were already comfortable and with which they already had an Imaginary identification. The project leaders encouraged this, as experience in the first year of the project (2004) had indicated, rather surprisingly to them, that adaptation of this kind was inevitable. However, there was an early move to discuss the theoretical underpinnings of MiC, with the intention of focusing prospective deviations toward a shared pedagogy, which, nonetheless, retained the characteristics of practices mandated in the new curriculum.

There is a tendency when discussing a project that involves a number of individuals to frame them all as a collective, and, thereby, gloss over specific responses. It is worth noting that the teachers here were volunteers and felt that they could be honest about their responses. On the Gatsby project, teachers did share concerns. There were issues around, for example, planning and preparation time, and class dynamics and management, including pace, the balance of discussion, independent working, written work, use of textbooks and ICT, and assessment, as well as other individualized concerns. Whereas each teacher shared at least some of these concerns, each response to MiC was enduringly individual, in terms of the degree to which teachers identified with particular concerns and how their perspectives changed over time.

We begin with Carol, who was one of a number of newly qualified teachers joining both the Gatsby and the ESRC project (2005–06). Carol worked in a school where, typically, student scores in national testing were above average and there was considerable pressure to increase this profile. The example below is concerned initially with the level of discussion in the classroom, but it has been selected here to illustrate the *Ché vuoi?* of teaching— the experience of trying to work out what to do—here sketched out by one teacher's attempts to identify and work with MiC approaches.

Experience as a student teacher at MMU had introduced Carol to discussion as a pedagogical tool bringing benefits to learners, and, at first, her strong identification with this approach as good practice is very evident. In practice situations, she had been well supported by mentors; however, as a newly qualified teacher, she had additional features of practice to bear in mind. Contemporary documentation had also alerted her to the need for pace, differentiation, and other features of a "good lesson," and here she ran into difficulties. At first, she endorsed discussion as a feature of good practice, and the endorsement carried the flavor of her training course:

> I think I like discussing with them because it enables the ones that are weaker and coming in with the low level 3s to access some of the maths that perhaps they can't write down, but they will perhaps have a go at discussing it. I tend to use a lot of discussion anyway in my lessons.

As the term progressed, difficulties began to accrue:

> Ok, the problem was, earlier on, I thought there was too much discussion going on. I felt like the kids were bored and I was bored and, looking back, it was because every single question was being answered, written down, then discussed and then discussed some more. So...I [thought] about differentiating again....Some of the kids working, doing more questions, and some not doing as many, but keeping them together and keeping the pace going. We went in today, so we talked through the first couple, the kids really got into the context, they were enjoying it. I let them go with it a little bit, then I pulled it back onto the problems. We talked about the first couple of questions, we moved on....I gave them a set time to do the written work, and that seemed to work a lot better, because we didn't then spend ages discussing the answers....That's what I was saying, talk about a question, discuss it, write the answer down, and talk about it some more, which was obviously not really the way we should have been doing it, which was discussed Friday [at a training day].

An important question for Carol is perhaps related to how long discussion ought to last. There is no clear answer here, as the necessity to perform discussion is rather better understood than what the performance ought to be about. While, at training sessions, there has been an attempt to deal with this by emphasizing the development of mathematical concepts, this has been privileged over other features of classroom life that intervene and make discussion problematic. Here, there is an issue of order, designated as "boredom," which must be attended to. The powerful discourses surrounding frameworks for teaching cause varied classroom difficulties to be pinned around particular sets of pedagogical solutions. In this instance, lack of "differentiation" and "pace" are identified as contributing to the problem. Although MiC advice suggests that pupils ought to stay together as a class, a decision is made to vary the emphasis placed on each question amongst the pupils and reduce the amount of discussion that takes place: "We talked about the first couple of questions and then moved on."

Although Carol has agreed to try out the new curriculum, she has not been freed from her obligations to the nationally imposed curriculum, the

National Numeracy Framework (NNF, 2003). She views each as distinct, and their varied approaches clash together around notions of pace. MiC does not lend itself to the notion of mathematics teaching and learning being "snappy" in ways that appear to be favored in the NNF; however, there is some assuaging of the difficulties arising, as MiC seems to offer learning in depth, a notion which Carol used to reassure herself:

> In an NNF lesson, it's more about kids getting the answers right, and less time is spent looking at the method than on the answers, and also that runs in line with less time spent with the context . . . than there is straight into the abstract, so I think it is less likely in an NNF lesson that you would discuss method, and more likely that you would just sit there and mark answers. Now that's something that I am trying to move away from, because I don't think it is worth doing at all . . . so I'm trying to go round and do MiC in NNF [lessons]. . . .

However, as the year progressed, Carol's concerns for her perceived difficulties with pace lingered:

> I think the problem with me was whether I was spending too long on questions initially, which I think I was. And I think, to an extent, I still think that. I think I am spending far too long on this. And I think it's difficult, you know, alright, you sit back at the end of a lesson and you think, right, ok, I spent too long on that, what am I going to do next time? . . . and the next time you spend too long on it and you think what am I going to do next time? . . . And I think sometimes it's nice to have a completely neutral view, saying this is how you could do it.

MiC favours both discussion and "moving on." Some of the other teachers acknowledged that, while understanding would always vary, there was a sense in which the perceived arbitrariness of the shift from one thing to the next worked against what they regarded as their better judgment. At such moments, there was often an invocation to have faith in the project. This is echoed in Carol's words here: "[W]hen I teach MiC, you're teaching them to do the same curriculum, in effect[, as the NNF]. You are teaching them the same things, but you're teaching them in such a way that it makes sense."

To return to theory, For Žižek (1989), Symbolic authority is always arbitrary and authoritative, rather than pointing to any "real" qualities that might be possessed by the individual. Yet, the individual feels called upon to demonstrate the qualities that are called for. Thus, the teacher believes in the necessity of demonstrating that discussion takes place in her classroom, but this functions as a performative move, rather than a pedagogical step that has made personal sense through experience.

Carol's story is characterized by her attempts to make her efforts "fit" the demands of the curriculum innovation, as she understands it. Her initial moves suggest that she had developed a comfortable identification with discussion as a desirable pedagogical tool, until the reality of her experience forced her to re-think this. Caught in a discursive framework that also privileged the frequently competing calls for a smart pace and differentiation, she felt the full impact of a position in which attention to one aspect of her mandate for development left her feeling vulnerable, in relation to others. Over the course of the year, there appeared to be little recognizable progress, at least in her understanding of discussion as realizable tool for learning. There is a strong sense of her to-ing and fro-ing between her sense of self as a teacher and the "rules," indicating a struggle to identify with the latter. Ultimately, she had managed to reconcile some of the competing demands made of her use of time, but the bewilderment associated with the question *Why am I what I'm supposed to be?* shows few signs of diminishing.

At one point, and knowing that it is fruitless, Carol petitions the Other, who, it is assumed, with a "completely neutral view," is in the know and can tell her exactly how it is all to be done. However, this perception of wholeness possessed by the master signifiers of the Other is a misrecognition, for the Other is also characterized by "lack in being."

This sense of oscillation, which marks the notional space of self/Other, occupied by the Imaginary, might well be anticipated as an area of conflict for NQTs; however, this is also the case for more experienced teachers in ways that echo Carol's experience while "working out what to do." But here, I raise a slightly different issue, which pre-occupied Marie, who, as an experienced teacher of some six years, practiced in an area of significant deprivation and identified closely with approaches fostering discussion in the classroom. She had already worked out a response to earlier models for teaching with which she identified closely, but was here "called" to do something different. The issue discussed below is the use of textbooks, the principle resource in MiC. Marie had used textbooks rather fitfully as additional materials prior to joining the Gatsby project, but, in general, preferred to find her own resources. Marie begins:

> And like the other day, I actually started off my lesson with "turn to page twelve, I want you to have a go at question nine . . ." and I would never do that in a lesson. "Turn to this page, do these questions," without any input: but the MiC actually lets you do that because they've done the arithmetic trees so they could just get on with it, but it just felt really wrong.

An initial reading suggests that Marie worked to carry out the mandate associated with MiC and experimented with the use of textbooks, "trying them out for size," as it were, or perhaps trying out her sense of identity as the teacher who uses textbooks. This trial was unsuccessful. Closely affiliated to practices which eschewed teaching from textbooks, Marie felt this was "wrong." Later, she appeared to change her mind:

> I try not to use textbooks religiously.... [E]very lesson is textbooks out, the kids know where they are with it, as well. They can follow it through and they know the structure of the lesson. We're going to get the books out and going to do whatever comes next in the books, so it is completely different from a normal lesson, I suppose, in that I just sort of dip into books when I need to....

A little later in the year, she acknowledged that an extended project would require everyone to work from the same text:

> If it was rolled out, for example, across even just schools in Manchester, you are going to have in every school a teacher that uses SMP, or Smile, or something like that, where it is an individual learning. It lends itself so well, because it explain as you go through and you've got your stories and you've got your context and kids can access it, that you could get the lazy teacher, "We're starting from page thirteen today, get on with it."

While, in the second extract, she appears to appreciate the orderliness that the materials bring and that this supports the pupils' learning (extracts 2 and 3), by the third extract, the material that lends itself to learning by providing "stories and "explanations" slides almost imperceptibly within the same sentence into the very material which supports the "lazy teacher." Nor are these and other difficulties ironed out as the project progresses. As Marie "catches herself" at one point relying more on the materials, she appears rather perplexed by her own behavior and its unpredictability:

> I have found, though, in my year eight and nine lessons, particularly now so, I'm relying more on the textbooks—and I've only just caught myself doing this recently, not using them every lesson, but certainly covering what's in the textbooks more than normally ... because you have to have faith in MiC work, it's making me have more faith in text books. But I don't like these textbooks. But yet, I find myself using them more and more.

Marie is rather surprised by her own "baffling" actions, where, despite not liking textbooks, she finds herself increasing her use of them. Additionally, having "caught [herself] doing this recently," the juxtaposition of having "faith in MiC work" and hating, yet using, their textbooks are conflicting notions that appear to pass unnoticed. To paraphrase Britzman (1998), the inside and outside are both highly complex. Yet the Imaginary is at work making us sound coherent to ourselves. In her final interview, Marie said:

I am back to the opinion of no, I don't want to use a textbook in other lessons, because it is stifling. So, yes, my opinions are changing all the time and I don't think, I know it's going to change probably back again on a different track, if you ask me next month.

Here, Marie conceded that, after a year, she cannot resolve the dilemma she faces, and like Carol, there is little sense of progress, at least in a conventional sense. On a conscious level, her identification as a teacher of mathematics *sans* textbooks appears to be too difficult to repress, although she acknowledges elsewhere that they had their uses. Marie's experience does not reflect the sense of subdued panic sometimes discernable in Carol's data and which gives it a certain feeling of immediacy. Though there is a strong sense that she, like Carol, is aware of the impossibility of satisfying the mandate she faces, she appears more prepared to live with the difficulties.

CONCLUDING COMMENTS

Educationalists, including teachers, share with policy makers a fantasy for classrooms where "an uncommon sociality can be invented in a common place" (Britzman, 1998, p. 50). In this fantasy, recommendations for practice are easy to follow and learners learn along predictable pathways and at a pace which is sufficient for them to acquire and secure the subject knowledge embedded in curricula, increasing their levels of attainment year on year.

The disposition of many teachers, their sense of vocation, and their general desire to be effective in their chosen profession makes them vulnerable to fantasies of this kind, despite the wealth of experience that reveals the many levels of difficulty involved in bringing improvement about. Fantasies enable us to function in the world and the Imaginary makes us feel that we act rationally in a world that we mis-recognize as already rational. Carol struggled to marry the orthodoxies of practice (instrumental, hierarchical, measurable) to classroom contexts where these features were difficult to fashion. However, the fantasies of policy makers provide a readymade conduit to the *objet petit a* of individual desires for practice, the practices which also appear to embody promise of improvement. Buying into fantasies of this kind cannot be avoided, as the individual cannot stand outside the Other, but we all respond differently, given our own preferences and inclinations. This aspect was much more visible in Marie's data as she struggled to resist textbooks and the poor practices that she, justifiably or unjustifiably, associated with them. Marie and Carol are representative of other teachers, insofar as they are different. Given the data emerging from the wider project, it is clear that models for development do not produce groups of teachers who necessarily share trajectories for development, despite centralized decision making (Brown et al., 2007), a feature which policy makers could usefully note.

The recognition that teacher identity is a feature immanent to change is an important one, and a psychoanalytic framework is a particularly apposite lens. While I have chosen to emphasize identity as a feature of the relationship between self/Other in a psychoanalytic framework, the fantasies and desires of the individual, their "loves" and "hates," are also hovering in the wings. Britzman (1998) suggests that these return to you in traces, underpinning attachments to some ideas and not others, contributing to the fragmentation of the self.

Žižek (1989) suggests that, as fragmented individuals, we cannot smooth out our identities to meet the goals laid out for us, as the goals themselves are embedded in conflicting discourses. Similarly, in the classroom, we cannot reconcile the conflicting demands set out for us into a smooth form of practice (Brown & McNamara, 2005). All this points to a need to recognize the individual teacher as an individual who has a sense of purpose that exists outside the lists of competencies designed to describe her. I will leave the final words to Britzman; there is always "something more than staying put in the logic of official knowledge" (1998, p. 49), and this needs to be valued.

Finally, I acknowledge the fantasies that I employ here to offer a more pertinent picture of the self/Other dynamic than I perceive in the work of policy makers.

REFERENCES

Ball, S. (2008). *The education debate*. Bristol: The Policy Press.

Britzman, D. (1998). *Lost subjects, contested objects: Toward a psychoanalytic inquiry of leaning*. Albany: State University of New York Press.

Brown, T., & McNamara, O. (2005). *New teacher identity and regulative government*. New York: Springer.

Brown, T., Hanley, U., Darby, S., & Calder, N. (2007). Teachers' conceptions of learning philosophies: Discussing context and contextualising discussion. *Journal of Mathematics Teacher Education, 10*(3), 183–200.

Coffield, F., Edward, S., Finlay, I., Hodgson, A., Spours, K., Steer, R., & Gregson, M. (2007). How policy impacts on practice and how practice does not impact upon policy. *British Educational Research Journal, 33*(5), 723–741.

DfES. (2002). *National Strategy, Framework for Key Stage 3, Mathematics*. London: DfES Publications.

Easthope, A. (1999). *The unconscious*. London: Routledge.

Fullan, M. (2003). *Change forces*. London: Falmer.

Lacan, J. (1977). *ECRITS* (Trans: A. Sheridan). London: Routledge.

Stronach, I., Corbin, B., McNamara, O., Stark, S., & Warne, T. (2002). Towards an uncertain politics of professionalism: Teacher and nurse identities in flux. *Journal of Education Policy, 17*(1), 109–138.

Van den Heuval-Panhuizen, M. (2000). Mathematics education in the Netherlands: A guided tour. *Freudenthal Institute Cd-rom for ICME9.* Utrecht, Utrecht University.

Wenger, E., McDermott, R., & Snyder, W. (2002). *Cultivating communities of practice.* Boston: Harvard Business School Press.

Žižek, S. (1989). *The sublime object of ideology.* London: Verso.

FURTHER READING

Barry, P. (2002). *Beginning theory.* Manchester: Manchester University Press.

Brown, T., Atkinson, D., & England, J. (2006) *Regulatory discourses in education.* Oxford: Peter Lang.

Brown, T., & McNamara, O. (2005). *New teacher identity and regulative government.* New York: Springer.

Butler, R. (2005). *Slavoj Žižek: Live theory.* London: Continuum.

Eagleton, T. (1996). *Literary theory.* Oxford: Blackwell.

Easthope, A. (1999). *The unconscious.* London: Routledge.

Hanley, U. (2007). Fantasies for teaching: Handling the paradoxes inherent in models of practice. *British Educational Research Journal, 33*(2) 253–272.

Homer, S. (2005). *Jacques Lacan.* Oxford: Routledge

Myers, T. (2003). *Slavoj Žižek.* Oxford: Routledge.

Sarup, M. (1992). *Jacques Lacan.* Hemel Hempstead: Harvester Wheatsheaf.

Torfing, J. (2003). *New theories of discourse, Laclau, Mouffe and Žižek.* Oxford: Blackwell.

MORE ADVANCED READING

Žižek. S. (1989). *The sublime object of ideology.* London: Verso.

CHAPTER 2

WHAT DOES IT MEAN TO CHARACTERIZE MATHEMATICS AS "MASCULINE"?

Bringing a Psychoanalytic Lens to Bear on the Teaching and Learning of Mathematics

Tamara Bibby

Masculinist ideologies idealise [the] logic of rational order, in which all things are subject to the imposition of an organising power, and contrasts it with the apparent chaos and unpredictability of emotionally driven disorder.
—Frosh, 2002a, p. 37

ABSTRACT

When we think about the social world, we have a tendency to categorize people, attitudes, and behaviors as either male or female, masculine or feminine

Unpacking Pedagogy: New Perspectives for Mathematics Classrooms, pages 21–41

(usually spoken in that order), and these labels come laden with powerful messages and expectations. Bronwyn Davies (1997) highlights the way these gender binaries have come to be seen as absolute, essential truths and facts, rather than stories we tell about ourselves and our world:

> [The Enlightenment] caught us up in the glorification of science with its domination of the rational (male) mind over (usually female) matter. In the scientific regime of truth there are two sexes. The differences between them are scientifically established as absolute, or essential. The two-sex model is one in which each sex takes its meaning in opposition to the other, any deviations are understood as aberrations, deviations from what *is*, and what *ought* to be. (p. 10)

It is precisely the difficulty of imagining beyond these binaries that keeps them in place and maintains their power. Indeed all dichotomies, not just the male–female binary, rely on the definition of one as "this," and, therefore, "not the other" (try defining heat without mentioning cold). Such definitions can exist only in opposition to their binary-pair; the essence of one is bound to a negation of the essence of the other. It is the destructive tit-for-tat of dichotomous thinking that locks us in conflict; in a world of *me* and *not me*, there can be no *us* (Benjamin, 2004). I suggest that the adoption and valorization of one side of a set of dichotomies and the concomitant rejection of the other side is an example of psychological splitting. I explore this more fully below.

If we accept for the moment the widely held belief that "mathematics is masculine," what implications does this have for teaching and learning mathematics? This chapter uses psychoanalytic theories to investigate this question. There are a number of ways the claim that mathematics is masculine might be read. The phrase elides distinctions that might be made, for example, between mathematics as a masculine subject (its content and processes) and mathematics teaching as masculine, or perhaps reproductive of particular masculinities. The version of masculinity that hovers over and is deemed to pervade mathematics relates to an unemotionally authoritative, rule-giving, rational, systematic, and logical set of values and practices, which simultaneously (and necessarily) separates itself from apparently feminine qualities, such as warmth, emotional attunement, and creativity (Bibby, 2002a; Shaw, 1995). The analysis developed in this chapter, while framed differently than Valerie Walkerdine's (1990, 1998) discussions of the role of gendered and classed discourses in the production of subjectivities, has strong similarities in intention. The conclusions demand a no less radical break with current practices.

Important to the argument developed in this chapter is a desire not to essentialize gender, that is, not to attribute masculine behaviors solely and necessarily to boys and men and feminine behaviors to girls and women (Butler, 1990; Youdell, 2006). In following these feminist, poststructuralist understandings of gender and psychoanalytic theories stemming from Klein and Riviere (1937), like Britzman (2009), I see masculinity and femininity, boy and girl, as elements within gender.

I am interested in people's relationships to school mathematics, rather than the subject *per se*, although separating these two things is never entirely possible. I tend to think about relationships to mathematics as relationships to authority. Britzman (2006) has written about pedagogical relationships and highlighted the many names that learning assumes in psychoanalytic writings (the Oedipus complex, the drive, infantile sexuality, insight) and in education (influence, authority, curiosity, affection). She points to the extreme difficulties of working with and between these terms. In this chapter I draw on psychoanalytic discussions of the oedipal family and the Oedipus complex to explore relationships to mathematics, particularly as they are constituted in primary schools. I will draw on post-Freudian psychoanalytic theories of authority to explore the sort of extreme masculinized notion of mathematics exemplified by traditional school mathematics pedagogy, with its emphasis on an adherence to rules, speed, and right answers. Drawing on a theoretical discussion, I suggest that a reconsideration of the experiences of learners of mathematics framed through the Oedipus complex can be used to recast the now-familiar argument that mathematics is masculine. I will then explore some potential consequences of the dichotomizing involved, the tensions that inhere in the situation, and the consequences for teachers and students living with the effects of these splits in policy and practice. Finally I will suggest a possible way to think beyond the dualisms and to see what it might mean, were education to "serve as gender's playground," although how do-able the suggestion is remains unexplored here (see also Gilbert, 2001; Walkerdine, 1998).

INTRODUCING IDEAS

This chapter draws on two psychoanalytic ideas: that of the defended subject (see Holloway & Jefferson, 2000) and the Oedipus complex (Freud, 1987).

The Defended Subject

Psychoanalysis suggests that we are all defended subjects, that we unconsciously protect ourselves from ideas and feelings we cannot bear. The defensive processes of splitting, projection, and denial are Kleininan concepts that describe a series of unconscious processes that we use to defend ourselves against ideas and emotions we find psychically painful, difficult, or otherwise unacceptable. In everyday parlance, these terms have strongly pejorative overtones; however, in a psychoanalytic context, they describe the human condition and carry no negative judgment—they are descriptions of ways of being, adopted and developed unconsciously. These processes have been described more fully elsewhere (e.g., Elliot, 2002; Frosh, 2002b), but, briefly, the suggestion is that aspects of ourselves that are experienced as unbearable are split off. We separate out our fear, anger, envy, vulnerability, for example, and deny their existence, projecting them onto others so that we can experience them as someone else's, removed from us. It is existentially safer to feel concern or contempt at someone else's terrifying vulnerability than to accept and experience our own. In terms of mathematics, feminine qualities, such as intuition or empathy, are the ones to be dreaded. The fantasy is that to allow any of these feminine qualities into the masculine world of mathematics would be to undermine and destroy it. This is the logic of the dichotomy: a single act of compromise would result in the death of mathematics, mathematical knowledge, and/or the mathematical self, so the qualities deemed feminine cannot be allowed to exist in mathematics at all and are projected entirely onto the arts and humanities. So, when the extreme "mathematical gaze" considers the arts, it feels the need to sneer at the childishness, the lack of rationality, the subjective nature of the judgments, and so on. Conversely, when the "arts gaze" considers mathematics, it shudders at and reviles the perceived lack of humanity in the cold, logical, rational order—a stance that is no less sneering. Sneering is defensive; it allows the maintenance of imagined psychic integrity. To educational researchers, this split is perhaps also familiar in the extreme positions taken by some quantitative and qualitative researchers. Although these processes are unconscious, they can have profound effects on us and those around us (see Frosh, 1999; Rustin, 1991).

The Oedipus Complex

The second element of psychoanalytic theory I draw on concerns the Oedipus complex. I realize that my desire to use this framing might seem strange since the Oedipus complex is the source of much humor and misunderstanding. However, the story, as it is used psychoanalytically, is richer than this and broader in its impact—indeed, for Freud, the Oedipus complex lies at the foundation of civilization, setting up the repressions and sublimations required by culture for its survival (Freud, 2001; Frosh, 1999). Initially, in its opening movements, it is about the intrusion of the father into the mother/child dyad and the child's entry into language and culture. Later, it is about the child's need to identify an object of desire—typically, although not necessarily, the parent of the opposite sex—and a rival—the parent of the same sex. At a social level, as Blackwell (2000) suggests, it can also be read as the site of a struggle over loyalties and allegiances within which desires must be negotiated. Furthermore, this negotiation of desires is driven not only by the individual's fantasies and fears, but it seems likely that the real seductiveness of one parent and real resentment are also highly significant. The argument developed in this chapter is concerned with the first parts of the story, the child's coming to know its parents as separate from itself, its negotiations of relations with each of them, and the kinds of knowledge and knowing that this entails and subsequently permits.

The psychoanalytic story begins with the new mother and her baby. The intense relationship is talked of as dyadic, the dreaming and associative mother attuned to her helpless baby and able to soothe his[1] distress through holding, feeding, and responding to tears and cries. At this stage, the baby does not differentiate itself from its mother; she occupies its field of vision and life, surviving its hunger, anger, frustration, and fury, and returning these difficult emotions in a form the baby can survive. The idealized (and therefore unattainable) image is of two beings in total harmony, blissfully satisfied with each other and their mutual needs and pleasures (see Winnicott, 1964, for further discussion).

At this point, theory suggests, the dyad is broken by the intrusion of the father, who declares that the two cannot continue this closed pattern of life[2]. Lacanian (French) psychoanalysis playfully changes the "Non" of the father (no, you cannot live forever united with your mother/child) to the "Nom" of the father, an allusion to "the name of the Father, the Son, and the Holy Ghost"—an appeal to a higher authority. Psychoanalytically, it is this prohibition that marks the separation of the child from its mother and its entry into the symbolic order of language; indeed, the other meaning of *nom* is "noun," a word for a thing that, in French, will be either masculine or feminine. Other theories might characterize this as the point at which the child starts to differentiate itself, to recognize self and other, *me* and *not me*, and to learn to stand beyond itself to reflect from other points of view. It is thought of as the beginning of ego differentiation, recognizing the existence of others beyond the self, and the development of the super ego, the mind's ability to absorb social laws and to self-regulate, to hear the *non* of the father without being told. The gendered nature of the roles of the caring, nurturing, empathic mother and the apparently dispassionate, law-giving father inform the dyadic gender split between soft femininity and hard masculinity identified by Davies (1997) and with which we are so familiar. The movement, framed by the Oedipus complex, is a developmental one, from wordless dependence to the rational mastery of language, culture, and an independent life. We read this as a "good thing," a "natural" and positive development, and that reading hides the heavy (psychic and material) costs of our acceptance (Walkerdine, 1990).

This is a story concerning very early childhood, so when considering oedipal influences in later relationships, it is important to make the leap to accepting that the story is metaphorical (indeed, it is based on Sophocles' non-existent, fictional family). This shift to accepting stories and concepts as metaphorical is always challenging, but lies at the heart of making sense of psychoanalytic theories. As adults, the roles or voices of the father, child, and mother are within us simultaneously and influence our relations with others at all times:

> Rather than emphasise the person of the mother or the father as objects to be internalised, I prefer to speak of them as bearing *orders:* sets of functions which engage and process the infant.... [I]t is important to bear in mind that these orders are not descriptions of how all mothers or fathers behave, but of processes associated with and usually conducted by the mother or father, who assume differing forms of significance for the developing infant and child. Behaviourally, the mother will perform paternal functions and the father will operate in the maternal order. (Bollas, 1999, pp. 37–8, emphasis in original)

Bollas is suggesting that speech and actions originating in the maternal or paternal orders may be expressed by either the mother or the father—that who utters them is not necessarily determined by the sex of the speaker. In a discussion that extends beyond family units, I would add that the maternal and paternal orders may be expressed by biological parents or other adults, and that such expressions are not confined to communications with very young children, but continue through the life course; oedipal relations—relations with, and development of, our own nurturance and authority—never stop. However, our pedagogic relations (as teachers and as learners) and our relations to knowledge and learning will depend, in large part, on the ways in which we negotiate this first complex dynamic and the demands made by the authority embedded within it. As someone interested in processes beyond the family, I transfer these *orders* to relations within a classroom and explore them in that context to see what light they can cast on my understanding of processes of learning and teaching. For the purposes of this chapter, then, we have a subject (school mathematics) which will be characterized in the next section as conforming to this notion of the prohibitive oedipal father (patermathematicus), a teacher who has internalized the law of the father from relations to her or his own father (paterfamilias) and of relations with and to other (in oedipal terms, inevitably masculine) authorities, and the child who, like the teacher, has relations to other authorities within and beyond the classroom, and who is in a different structural power-position.

APPLYING IDEAS

Although various epistemologies of mathematics exist, they can loosely be characterized as being concerned either with *mathematics as a product* (an absolutist notion of a mathematics that exists "out there," waiting to be discovered) or *mathematics as a process* (mathematics as a result of human endeavor, a socio-cultural artifact). Of these, the dominant belief is that mathematics is a product, written in the stars and awaiting discovery. An example of the absolutist belief in mathematics as a naturally occurring product can be read through English journalist Melanie Phillips' complaint about what she perceives to be a crisis in the teaching of mathematics in English schools:

> The subject that has suffered perhaps most unequivocally from the collapse of the authority of rules is mathematics. Maths teaching in Britain has effectively been deconstructed. A fundamental shift in emphasis away from knowledge transmitted by the teacher to skills and processes "discovered" by the child has undermined the fundamental premises of mathematics itself. The absolutes of exactness and proof on which the subject is based have been replaced by approximation, guesswork and context. . . . The emphasis on the practical application of maths and the obsession with presenting numbers "in context"—in other words, relating them to real life situations—[has] denigrated the primary importance of maths as a training for the mind. (Phillips, 1996, pp. 13–14)

Here, we have mathematics characterized as the perfect example of the "authority of rules," a subject that is quintessentially knowable, teachable, and testable. It is exact, precise and rational, "invested in reason's dream of a calculable universe: the control over time and space" (Walkerdine, 1998, p. 165). Phillips contrasts this pristine cleanliness with the mess of "approximation, guesswork and context," the inexact, imprecise spongy warmth of contingent relationships and chaotic bodies. Indeed, it is notable that among the topics she uses as a stick with which to lambaste "trendy teaching" is the "analysis of HIV/AIDS statistics (essential no doubt to every child)" (1996, p. 43): a fine proxy for the terror of unknown "dirty" bodies that need to be kept away from the purity of childhood. It is against the powerful background of this prohibitive notion of mathematics that the discussion here takes place.

Mathematics as Masculine Reconsidered

> [W]ithin current school mathematics practices, certain fantasies, fears and desires invest "man" with omnipotent control of a calculable universe, which at the same time covers a desperate fear of the desire of the Other, "woman." "Woman" becomes the repository of all the dangers displaced from the child, itself "father" to the man. (Walkerdine, 1990, p. 74)

As noted earlier, masculinity (or, more specifically, prohibitive oedipal masculinity) can be characterized by the qualities in one side of a list of dichotomies, such as those in Table 2.1.

TABLE 2.1   Masculine/Feminine Dichotomies

Masculine	Feminine
rational	emotional
ordered	creative
rule based	emotional
numbers	words
abstract	concrete
competitive	collaborative
independent	dependent

When people talk about why they didn't like mathematics in school, they tell us about its cold rationality, about the importance of using "the right" method, about following rules, the lack of room for an idea or an opinion, about the fear and humiliation of being wrong, and the lack of place for self (Bibby, 1999); they express a yearning for a different experience with a less censorious mathematics (Brown et al., 2008). Of course, some find the certainty they see mathematics providing as comforting and dislike having to express an opinion or explain themselves; they reject what they see as the subjective contingency of more "feminine" subjects. In these positionings, we hear enacted the split experienced between the arts and the mathematical sciences (mathematics, physics, chemistry, etc). All the creativity and emotionality, all the relationships and relatedness are located in the humanities while rules, certainty, logic, and rationality are emptied into mathematics and the sciences, with the two held separate (Harding, 1991; Shaw, 1995). Coming to experience oneself in mathematics lessons and learning, through the judgments of mathematics teachers and the ticks and crosses in an exercise book, to accept a set of painful stories about oneself as a learner are things many people talk of (Bibby, 2002b). As suggested earlier, operating in a dichotomized system makes it very difficult to refuse to be positioned on one side of the divide or the other, we are expected to embrace, and find a home within, the organizing dichotomies. More often than not, to stand in one is to deny the other, and only rare creatures are allowed publicly to occupy both. You can be an "arty" person or a "geek," but not both.

For those like Melanie Phillips, who believe the curriculum needs to be based in masculinist, product notions of mathematics, there is a double necessity to simultaneously lament the feminization of mathematics (softening, lowering of standards, its "reduction" to applications, and its miring in the messy business of life), as well as to man the barricades to defend their own stance (the gold-standard, trainer of the mind, developer or mark of "raw" intelligence). Finding some other place to stand seems problematic, faced with these trenchant and entrenched voices.

The dichotomized idea of mathematics as masculine fulfills the prohibitive oedipal function of the father by forbidding contact with the maternal order—the taboo keeps mathematics and the humanities apart. Furthermore, because we are projecting our fantasies of the father onto mathematics, imbuing a non-sentient subject with masculine qualities, it cannot have the mitigating qualities of any human male. There are no residual connections with, or experiences of, an internal or external mother. This uber-masculinist creation is, by definition, totally denying of any feminine qualities, and our relations to it place us (women and men alike) in an impossible position. To be loved (and accepted) by mathematics, the ultimate rational and prohibitive father, will be difficult. This is not to say that there are not those who form a different relation to mathematics (or find a different mathematics to relate to), or who manage to come to some accommodation or even enjoy the relationship, but it is to express a lack of surprise that, for so many, the relationship is experienced as abusive.

Dichotomies such as that between the mathematical sciences and humanities demonstrate a splitting, operating at a cultural, as well as individual level, and suggest that, to some extent, we—individually and societally—need to keep the maternal and paternal orders separate, that there is something painful in the mix against which we defend. In the context of a psychosocial exploration, important questions are: What does the splitting/dichotomizing defend against? What purpose does it serve?

Policies for Teaching

When the National Numeracy Project (now the Strategy) was first piloted in English schools, the influence of academic mathematics educators could be read clearly in the associated texts (for example, National Numeracy Project, 1996; Qualifications and Curriculum Authority & The National Numeracy Strategy, 1999). In the early days of this new policy incursion, through membership of advisory panels and writing teams, this group of "process-oriented" mathematics educators influenced both the form and content of the policy outputs. Within the framework of bounded lessons, of *should knows* and *will learns*, were softer voices extolling children's methods, exhorting teachers to work with the inchoate ideas of the classroom to develop mathematical thinking (DFEE, 1998). This was talk from a different, softer place: Listen to the muddle and the ideas and help this develop; explore together; help children come to a realization of inconsistency. In its movement towards a masculine ideal without an accompanying denial of the maternal order, it lacked the stridency of the patermathematicus.

The resulting tensions generated by the apparently contradictory messages about mathematics were felt by everyone and, for most, the option of a "third way" was not trusted to be real: There is no room for a third in a dichotomy. Later research looking at the ways in which the Strategy was being implemented in schools (Askew et al., 2002) showed clearly that, in the face of this set of contradictions, most teachers were retreating to the voice of the prohibitive maths-father; at least the maths-father clarified where one is to stand and what one is to say and do. The "other" softer, more nurturing voice was not simply being passively ignored; it was actively rejected, pushed away, denied. The dichotomy was being reasserted. So, the efforts of those who would widen the impoverished mathematical diets of children in schools were rejected. Teachers (and children) fled from the potential for strategic choosing. The terror of making a wrong choice and being punished for it was too great (and perhaps too real in the current climate in schools). Most turned with a sigh of relief and ran back to the prohibitive, oedipal father, who simplifies our lives for us, reducing the risk of learning and fear of failure. The split, then, defends against uncertainty and the difficulties of learning, and especially of learning from experience (see Bion, 1961).

Teaching Mathematics Reframed

I want to return to the dichotomizing of the curriculum, particularly the mathematics/humanities split, and the need to stand in one place or the other (whether as a result of the love of an embrace or the relief of a rejection). Why might it be so difficult to be a process-oriented, "connectionist" teacher of mathematics? And, are we inevitably trapped by the choices we consider? Are there only the oedipal maternal and paternal orders? How can there be anything else?

Perhaps surprisingly, within such a dichotomized system, the Effective Teachers of Numeracy project (Askew et al., 1997) developed a three-part typology of primary teachers' beliefs about, and orientations towards, mathematics and its teaching. While the authors were careful to state that individual teachers are a mélange of attitudes and dispositions, the caricatures of the typology are useful for our purposes. Although the *transmission* and *discovery* orientations can be placed easily within the oedipal framing, the *connectionist* orientation requires more thought.

It is not difficult to recognize a transmission orientation as coming from the paternal order, and the difficulties it raises relate to its denial of the maternal order. This is the vignette used to illustrate the practices of a teacher coming from a transmission orientation:

> A class of Y4 pupils are working on equivalent fractions. The teacher draws a diagram on the board to demonstrate a means of converting ½ into quarters.

She explains that the quarters are the fraction to convert to and so the pupils will need to draw a rectangle divided into four equal parts. Since half is required then two of these parts need to be shaded in. "So, half is equivalent to two quarters," explains the teacher. "On the other hand," she continues, "we could just look at the numbers on the bottom of the fraction. I have to multiply 2 (pointing to the 2 on the bottom of the ½) by 2 to make 4 (pointing to the four on the board of a yet denominator free quarter fraction), so I multiply the 1 (pointing to the one on the top of the ½) by 2 also. So we get 2/4, which is the same as we got when we drew the diagram." The pupils are given a number of fractions to convert into equivalents and told they can either use the diagram method or the multiplication method. As the teacher moves around the class, once pupils have done a few examples using the diagram, she suggests to them that it will be quicker to use the other method. (Askew et al., 1997, p. 26)

The teacher encouraged the children to move from the preverbal (drawing) into the method she perceived to be more mathematical (numerical) as soon as possible. While the numerical method may often be the quickest way to deal with equivalent fractions, it is conceptually less easy to discern what is being transformed. She appears to be concerned that the children do not spend too long developing understanding, and instead move to a performance of the more abstract form. We could read this as the teacher trying to protect the children from what might be seen to be the chaos and relative clumsiness of pictorial solutions in mathematical contexts, or her avoidance of a more nurturing, attuned way of being with the children (or both). The seemingly more mathematical method has been presented as the rational choice, a series of "logical" steps that are to be preferred. This kind of rationalization devalues the work of those children who prefer to progress more slowly, perhaps those who derive pleasure from the drawing, or those who are unable, at that point, to conceptually connect the two representations.

A discovery orientation is also portrayed with a vignette. Here, we might interpret the teacher as operating within the maternal order. The consequent conflict with the researcher relates to the teacher's denial of (and the researcher's acceptance of) contact with the paternal order:

The teacher is working with a low attaining [Y2] group on doubling. The pupils spend a long time counting out individual cubes, fitting them together and recounting them. The teacher sets them a series of numbers to double, reiterates how to find the answer using the cubes and goes on to join another group. [In conversation with the researcher...the] pupils are able to talk about what double four hundred might be and quickly move onto discussing doubling three thousand, six million and so on. After the lesson the teacher explained that she was concerned that the pupils were not ready to be working with such large numbers, particularly as no Dienes blocks had been got out so that they could see that double three thousand was six thousand. (p. 27)

In this example, the mathematics is assumed to be located within the blocks, the mathematical bodies being manipulated on the table tops. Any move to numbers as abstract concepts is being resisted by the teacher. The children are construed as not being ready for this, although, with the researcher, they are clearly able to engage with numbers as abstract mental constructs. It is not unreasonable to speculate that the teacher may have experienced mathematics classrooms negatively during her own schooling, that their disconnection from the ways in which she otherwise experienced herself as a knower caused her to defend herself against the painful experiences. She may be anxious that the cold, abstract rationality of mathematics might again make her lose any positive sense of herself as a knower (see also Povey, 1997, for further discussion of this phenomenon). Understood like this, as a professional, the teacher may experience her own anxieties about mathematics as unacceptable. We might suggest that she appears to be simultaneously projecting her anxieties about the abstract nature of numbers onto the children, so she believes it is they who cannot manage to survive contact with abstract mathematics, rather than herself (Britzman & Pitt, 1996). If this is the case, it is unsurprising that she may feel the need to protect them (from her own anxieties) by keeping the mathematics concrete.

A Psychoanalytic Reframing of the Problem

Any learning is potentially psychically dangerous. It involves taking a risk. To learn, we must expose ourselves both to the possibility of discomfort, failure, negative judgment, and, ironically, also to the dangers inherent in success: raised expectation, increased opportunities for failure, the envy of others, and profound loss. It can be hard to admit that there may be loss in learning; we are so used to celebrating it as an unalloyed achievement that the loss of self-who-does-not-know is overlooked. Jonathan Silin (2006) provides us with a poignant example from his own experience as a teacher of young children:

Along with parents I celebrated familiar milestones—a child learns to button her coat, tie her shoelaces, or walk home from school on her own. I did not imagine that in learning a new way of being in the world, a child might also give up an old way, one that had worked for him in the past—the physical intimacy that occurs when an adult cares for his clothing or the social connection he experiences when accompanied by a caregiver on the walk home from school. (p. 233)

As Bion suggests, in life "there is a continuous decision to be made as to whether to evade pain or to tolerate and thus modify it" (Symington & Symington, 1996, p. 28), and learning can only happen if the pain is engaged with. The pleasure that accompanies success (whether validated internally or externally) has to be enough to mitigate the risk, to carry the learner past the fear, with its accompanying inertia or flight. For many, the context of mathematics provides the one place where this risk cannot be borne. Paradoxically, generalist primary teachers are legally and professionally obliged to find a way past this (Bibby, 2002b). They are caught up in a Catch 22 situation: To contain their own anxieties around mathematics means fleeing into the maternal (or paternal) order. However, to successfully enable children to learn requires that they be able to contain the children's anxieties around mathematics. They need to develop an ability to "hear" that to which (in an effort to protect themselves) they have deafened themselves. Defenses always develop for a reason, although they can outlive their usefulness, becoming counter-productive as we continue through life. However, changing them is never easy, requiring a very personal kind of learning/ risk taking. It is not enough just to contain the anxious child (internally or externally); they also need to be challenged appropriately. Simply removing the risk and making everything okay for learners also removes opportunities to learn. Indeed, Britzman (2003) reminds us of Freud's warning to teachers that "idealizing the world for children and promising them happiness in a life without conflict would only incur helplessness and future disappointment" (p. 2). Many teachers do not "contain" a learner's fears and anxieties. Through their micro-management of choices and activities, they come to deny the fears, stopping the child from learning that the fears are survivable, and they do this whichever side of the split with which they operate. As teachers, we need to be brave enough to enable the student to be brave enough to risk learning.

RE-ENGAGING WITH THEORY TO MOVE FORWARD

There were attempts during the 1970s and '80s to develop girl- or woman-friendly mathematics programs (for example, Buerk, 1982; Becker 1995, Jones & Smart, 1995). These projects represented a range of initiatives, including work with older women on re-authoring their relationships with mathematics, to attempts to contextualize mathematics problems in more "feminine" contexts (e.g., cooking, rather than sports). (See Walkerdine, 1998, for an extended critique of these movements). Upon reflection, two things appear to have happened as a result of these initiatives: First, the programs may not have been particularly successful, in that they did not increase the numbers of girls and women continuing mathematics in post-compulsory education. This suggests that, while they may have been beneficial for individuals, they did nothing to change underlying beliefs and experiences of the nature of mathematics as a masculine subject. Second, they contributed to a backlash (Faludi, 1999), with, in the UK, increasing levels of policy prescription and a reassertion of the masculinity of mathematics.

It seems clear that the difficulties for mathematics are the notion that it is a masculine subject and the simultaneous need to accept its masculinity. We are trapped in the force field of dominant, apparently "natural" masculinities and the associated need to reject the over-absorbing, nurturing mother/femininity. In terms of teaching mathematics, this might be thought of as the balance required to both hold onto the fact that mathematics *is*, in part, at least, about rules, precision, rational logic, and so on, while not denying its creativity, its inexactness, its need for collaborative endeavor or the leaps of imagination required. It is about the simultaneous need for dependence and independence. It is about allowing these apparently contradictory states of being and ways of knowing to coexist and be mutually constitutive: an unsettling demand. Stephen Frosh (2002a), drawing on Jessica Benjamin, seeks to reframe the difficulty and to move beyond the dichotomies. He suggests that the strength and power of the oedipal father presents particular difficulties, in that it denies the male child any possibility of forming a loving relationship with his father: "[H]is failure to find a father who is present in any form other than as a symbolic prohibition, who is really there for the boy, who is in the positive sense "embodied." The splitting entailed in this leaves everyone impoverished" (p. 38). Again, we are faced with the difficulty of reading this analysis metaphorically. We might transpose it to mathematics pedagogy by noticing how hard it is for students to engage with mathematics as it is generally taught currently. And while mathematics can never be embodied in the same way a flesh-and-blood father can, we might think of the teacher and pedagogic relationship as providing the embodiment referred to. In claims familiar from other places, Brown et al. (2008) quote students who fervently declare that they would "rather die" than continue with post-compulsory mathematics study because "it SUCKS and I wouldn't want to spend any more of my time looking at algebra and other crap" (p. 10). In providing a thin form of mathematics, the prohibitive patermathematicus acts through the teacher to impoverish the learners' understandings and possible futures. In pointing out and lamenting the loss or concealment of a present father (a connectionist, process-oriented mathematics), Frosh, with Benjamin, Kristeva and others, points to a difficulty with the Oedipus complex:

> And so it was no accident that Freud's theory gave the woman a son to love, in his view the only unambivalent love. To the mother was granted the fulfillment of the wish for identificatory love not just in relation to her father but in

ideal love for her son; to the son was given the grandiosity that was mirrored by the mother who renounced it. But Freud's story also gave the son an ideal love, a forever unrequited love of the father who cast him out as a murderous rival. (Benjamin, 1995, p. 114, cited in Frosh, 2002a, p. 38)

For the son who would love his absent father and finds himself cast out, we can think about the children who would engage with mathematics, but whose challenges to mathematical authority are doomed to dismissal. After all, mathematics is about right and wrong answers, and school children are in no position to challenge the mathematics canon—indeed, few can. It is hard to turn away from unrequited love; there lingers the hope that one day, if only one can be the right kind of person, the desire for emotional closeness will be fulfilled. Unfortunately, because the demand (the definition of "the right kind of person") is beyond one's control, this is a fantasy space and all such yearning is doomed to remain forever unfulfilled (for a fuller discussion of the dynamics of emotionally abusive relationships see, for example, Shaw, 2005). And so it is that so many find themselves unable to live up to the mathematical ideas that are held up: the bright shining minds, the insightful clarity of logic, the speed and accuracy, and the disconnected cold rationality. In mathematics lessons, many find themselves less than perfect, subject to frustrations and furies as much as to pleasures and satisfactions and give up the dream: "I'm just not clever enough" or "I'm not that kind of person."

While this prohibitive oedipal father might not be the only one, can another (or a different, connectionist version of mathematics) exist as anything other than fantasy? Is prohibitive, oedipal masculinity the only model available in psychoanalytic theories? To begin to answer this question, we need to return to the original dyad: the dreaming, associative mother with her very new child. This dyad can also be thought of as potentially suffocating: In thinking the child's thoughts and taking care of its needs almost telepathically, the mother can be seen to be engulfing the child, holding it in stasis and halting its growth to prevent the loss of the object of her love. In this relationship, the child has no room (no need) to think and to learn. The example of the Year 2 class doubling only small numbers because there was no visual evidence of the larger ones stands here. There are two ways to imagine this engulfment and obsessive absorption ending. One, the traditional Freudian solution, is the intrusion of the prohibitive oedipal father, who, in standing between the mother and her baby and refusing their close contact, provides the child with space and the need to enter language and culture—the new teacher who will shake the class out of their dependence on their teachers/helpers/mothers and "drill" or "drum" (and other violent and intrusive metaphors) the skills and knowledge in. However, others have suggested another possibility in the turning of the mother away from her engulfment with her child—to look again to the man she loves, the pre-oedipal father-to-be. This turn is her choice and not reliant on the interrupting "No" of the father, but it, too, provides the child with a space into which she or he can grow: "The loving mother, different from the caring and clinging mother, is someone who has an object of desire; beyond that, she has an Other with relation to whom the child will serve as a go-between" (Kristeva, 1983, cited in Frosh, 1999, p. 232). For the mother (and the father), the father exists in two states: as her lover, and as the provider of the "*Non*" in their oedipal relationships with the growing infant. Interestingly, in Kristeva's formulation, the child has an active role, and participates in and is part of the relationship between its parents. This offers the possibility of a student who can act within the relationship between a teacher and her mathematical understandings, and it offers the possibility that the teacher (and mathematics) could learn from a child's engagement with a mathematics lesson. The difficulty is that Western masculinity is built upon the prohibitive oedipal father, and the pre-oedipal one, the loving man, gets forgotten/abjected. It is worth quoting Frosh (2002a) at length:

> Psychoanalysis also suggests that masculine gender identity is built largely on the repudiation of precisely those characteristics which might make this loving fathering possible: the boy child is thought to turn away from the mother and all she represents (dependency, intimacy, bodily absorption) in order to carve out a separate existence in the shadow of the absent father. Too much closeness to the maternal leads to the dissolution of masculine identity; moreover, the more fragile the sources of this identity . . . the more vivid and vicious this repudiation can become. Hence the specific complexity of the ambivalence surrounding fathering: to father a child requires something other than the traditional boundary-setting and prohibitive stance, no authority is vested in the father to sustain that stance; but to reach out in a loving way requires a shift of masculine consciousness involving not just some gentleness, but a whole gamut of alterations in relations of dependency, intimacy, vulnerability and trust. On the whole, the more fragile masculinity becomes, the more desperately men cling to its vestiges, doing the opposite of what fathering requires. (p. 33)

Herein, then, is the challenge for mathematics education, for teachers and learners at all phases of education. We need to rediscover the loving and nurturing pre-oedipal father and give mathematics a chance to become different. This is not a simple "feminization" of mathematics, although it may be ridiculed as such by those, like Melanie Phillips, who are terrified of other possibilities and split to defend against the challenge to their psychic safety. Indeed, to simply (or not so simply) try to construct a new, "girl-friendly" mathematics is to maintain the split cast by the oedipal order. Importantly, nor is it a denial of the fact that mathematics is masculine. Many of the qualities of mathematics *are* those "masculine" qualities of reason, logic, and so on. But it is to question whether the way we currently define and practice school mathematics (the curriculum, the examination structure, and the pedagogy) is the extent of what it can be. Bollas (1999) highlights the danger of allowing dichotomous thinking to dominate:

> No single one of these authors [the pre-verbal child; the dreaming, associative mother; or the paternal interpreter] will establish an ultimate truth; indeed, as time passes, each form of knowledge finds itself in a family of truth seekers and narrators. . . . If one of the three members of the triad . . . becomes too influential, or, if one function is eliminated completely, then full knowing is not possible. (Bollas, 1999, p. 38)

And yet, this elimination would seem to be what happens in school mathematics lessons where the pre-oedipal orders—and particularly the pre-oedipal, loving father—is entirely denied, sacrificed to the "*Non*" of the oedipal father. What I am suggesting is not some "third way"—not yet another state which would idealize and dream the split oedipal maternal and paternal orders away, but rather the hard work of a move to some unsplit (or less split) state in which we (and mathematics) are able to tolerate and hold the contradictions and tensions that exist between the maternal and paternal orders—something that will be particularly challenging for the fragile masculinities within all of us.

How do we reframe this in the context of teaching? Silin (2006, p. 236) draws on Ellsworth, Grumet, and others to explore the difficulty of pedagogy that, despite the fantasies of education and education policy, is "unpredictable, incomplete, and immeasurable in its impact." Working between Grumet and Sedgwick, he develops the suggestion that it is the teacher's responsibility to "point to part of the world that matters to her," identifying "a problem or experience worthy of the student's attention." In this act, the emphasis is on "the value of pointing and hinting over naming and telling." This is an invitation to the learner, not an exhortation. It is neither the prohibitive oedipal exhortation to leave nurturance and safety and to enter the drafty corridors of culture, nor the suffocating exhortation to stay forever joined in a state of suspended dreaming. It is precisely the difficult invitation to enter an uncertain space filled with the certainties and uncertainties of others and to carve and create meaningful knowledge from the possibilities that surround and suffuse us—to learn to survive our own fears and anxieties, and so to learn.

It is unfortunate that Silin leaves the problem solely with a female teacher. As Walkerdine (1998) reminds us, "the problem is not teachers . . . they are ensnared too" (p. 166). We need to remember their locatedness within the same discursive, historical, and cultural frames that allow others, in a tit-for-tat move, to cast the blame and avoid the taint of responsibility.

Teachers are complicit—as we all are—in the maintenance of the fiction of mathematics as masculine. But Silin's suggestion that what we need is less naming and telling, and more pointing and hinting, seems important. It is a suggestion that might usefully be thought about by all in education, from teachers and students in classrooms to government ministers and the civil servants who draft policy and assessments. It is also a suggestion that could have the potential to radically change mathematics and its teaching.

The theoretical psychoanalytic frame I have employed, while not perfect, and bringing difficulties of its own, does provide a different set of metaphors with which we could begin to think again about this very old problem. But there is no simple move, no simple solution on offer, just the hard, painful, and ongoing work of reworking and rethinking our relationships and actions within them. It will still remain for children/learners to come to know their desires and to form loyalties and allegiances within their peer and pedagogic relationships, but we might be able to provide less inevitable choices than those that seem to exist currently.

DEDICATION

This chapter is dedicated to the memory of Eibhlin Breathnach 1947–2005.

NOTES

1. The baby is always male. See Chodorow (1999) for a discussion of this.
2. Other psychoanalytic traditions place equal or greater emphasis on the mother's role in weaning the child, particularly the less literal weaning away from dependency with her and her gradual facilitating of a move towards the world beyond the dyad; that is, to the father and then the outside world. See, for example, Winnicott (1964).
3. Again, Winnicott (1971, see p. 18) would approach this differently, and use weaning as the metaphor through which to inform his discussion, although the result of the reading would be largely the same.

REFERENCES

Askew, M., Brown, M., Rhodes, V., Johnson, D., & Wiliam, D. (1997). *Effective teachers of numeracy* (Final report). London: King's College.

Askew, M., Bibby, T., Brown, M., & Hodgen, J. (2002). *Mental calculation: Interpretations and implementation.* (Final report). London: Department of Education and Professional Studies, King's College.

Becker, J. R. (1995). Women's ways of knowing in mathematics. In P. Rogers & G. Kaiser (Eds.), *Equity in mathematics education: Influences of feminism and culture* (pp. 163–174). London: Falmer Press.

Benjamin, J. (2004). Beyond doer and done to: An intersubjective view of thirdness. *Psychoanalytic Quarterly, LXXIII,* 5–46.

Bibby, T. (1999). Subject knowledge, personal history and professional change. *Teacher Development, 3*(2), 219–232.

Bibby, T. (2002a). Creativity and logic in primary school mathematics: A view from the classroom. *For the Learning of Mathematics, 22*(3), 10–13.

Bibby, T. (2002b). Shame: An emotional response to doing mathematics as an adult and a teacher. *British Educational Research Journal, 28*(5), 705–722.

Bion, W. R. (1961). *Experiences in groups and other papers* (2004 ed.). Hove, East Sussex: Brunner-Routledge.

Blackwell, D. (2000). The psyche and the system. In D. Brown & L. Zinkin (Eds.), *The psyche and the social world: developments in group-analytic thinking* (pp. 27–46). London: Jessica Kingsley Publishers.

Bollas, C. (1999). *The mystery of things.* London: Routledge.

Britzman, D. P. (2003). *After-Education: Anna Freud, Melanie Klein, and psychoanalytic histories of learning.* Albany: State University of New York Press.

Britzman, D. P. (2006). Sigmund Freud, Melanie Klein, and Little Oedipus: On the pleasures and disappointments of sexual enlightenment. In G. M. Boldt & P. M. Salvio (Eds.), *Love's return: Psychoanalytic essays on childhood, teaching and learning* (pp. 165–184). New York: Routledge.

Britzman, D. P. (2009). The madness of lecturing on gender. *Keynote lecture presented at the Gender and Education Association 7th International Conference. March 25–27.* Institute of Education: London.

Britzman, D. P., & Pitt, A. J. (1996). Pedagogy and transference: Casting the past of learning into the presence of teaching. *Theory Into Practice, 35*(2), 117–123.

Brown, M., Brown, P., & Bibby, T. (2008). 'I would rather die': Reasons given by 16-year-olds for not continuing their study of mathematics. *Research in Mathematics Education, 10*(1), 3–18.

Buerk, D. (1982). An experience with some able women who avoid mathematics. *For The Learning of Mathematics, 3*(2), 19–24.

Butler, J. (1990). *Gender trouble: Feminism and the subversion of identity.* London: Routledge.

Chodorow N. J. (1999). *The reproduction of mothering.* Berkeley: University of California Press.

Davies, B. (1997). Constructing and deconstructing masculinities through critical literacy. *Gender and Education, 9*(1), 9–30.

DFEE. (1998). *National numeracy strategy: A framework for mathematics.* London: HMSO. (But see earlier editions, e.g., November, 1998 for a greater variety of calculation strategies)

Elliott, A. (2002). *Psychoanalytic theory: An introduction.* London: Palgrave.

Faludi, S. (1999) *Stiffed: The betrayal of modern man.* London: Chatto and Windus.

Freud, S. (1987). The development of the libido and the sexual organisations. In J. Strachey & A. Richards (Eds.), *Introductory lectures on psychoanalysis* (Vol. 1, pp. 362–382). Harmondsworth: Penguin. (Original work published 1917)

Freud, S. (2001). Civilization and its discontents. In J. Strachey & A. Freud (Eds.), *The standard edition of the complete psychological works of Sigmund Freud. Volume XXI* (pp. 59–145). London: Vintage, Hogarth Press. (Original work published 1930)

Frosh, S. (1999). *The politics of psychoanalysis: An introduction to Freudian and post-Freudian theory* (2nd ed.). London: Macmillan.

Frosh, S. (2002a). *After words: The personal in gender, culture and psychotherapy.* Basingstoke: Palgrave.

Frosh, S. (2002b). *Key concepts in psychoanalysis.* London: The British Library.

Gilbert, J. (2001). Science and its 'Other': Looking underneath 'woman' and 'science' for new directions in research on gender and science education. *Gender and Education, 13*(3), 291–305.

Harding, S. (1991). *Whose science? Whose knowledge? Thinking from women's lives.* Milton Keynes: Open University Press.

Holloway, W., & Jefferson, T. (2000). *Doing qualitative research differently: Free association, narrative and interview method.* London: Sage.

Jones, L., & Smart, T. (1995). Confidence and mathematics: A gender issue? *Gender and Education, 7*(2), 157–166.

Klein, M., & Riviere, J. (1937). *Love, hate and reparation: Two lectures.* London: Hogarth Press.

National Numeracy Project. (1996). *Course reader.* Unpublished handout from initial training for consultants.

Phillips, M. (1996). *All must have prizes* (1997 ed.). London: Little, Brown and Company.

Povey, H. (1997). Beginning mathematics teachers' ways of knowing: The link with working for emancipatory change. *Curriculum Studies, 5*(3), 329–342.

Qualifications and Curriculum Authority & The National Numeracy Strategy. (1999). *Teaching mental calculation strategies: Guidance for teachers at key stages 1 and 2.* London: Qualifications and Curriculum Authority.

Rustin, M. (1991). *The good society and the inner world: Psychoanalysis, politics and culture.* London, New York: Verso.

Shaw, J. (1995). *Education, gender and anxiety.* London: Taylor Francis.

Shaw, J. (2005). Lacanian demand and the tactics of emotional abuse. *Psychoanalysis, Culture & Society, 10*(2), 186–196.

Silin, J. (2006). Reading, writing and the wrath of my father. In G. M. Boldt & P. M. Salvio (Eds.), *Love's return: Psychoanalytic essays on childhood, teaching and learning* (pp. 227–242). New York: Routledge.

Symington, J., & Symington, N. (1996). *The clinical thinking of Wilfred Bion.* London: Routledge.

Walkerdine, V. (1990). *School girl fictions.* London: Verso.

Walkerdine, V. (1998/2005). *Counting girls out: Girls and mathematics* London: Falmer Press.

Winnicott, D. W. (1964). *The child, the family and the outside world.* London: Penguin.

Winnicott, D. W. (1971). *Playing and reality.* London: Routledge.

Youdell, D. *Impossible bodies, impossible selves: Exclusions and student subjectivities.* Dordrecht: The Netherlands.

FURTHER READING

Bibby, T. (2009). How do pedagogic practices impact on learner identities in mathematics? A psychoanalytically framed response. In L. Black, H. Mendick, & Y. Solomon (Eds.), *Mathematical Relationships in Education: Identities and Participation* (pp. 123–135). London: Routledge.

Boldt, G. M., & Salvio, P. M. (Eds.). (2006). *Love's return: Psychoanalytic essays on childhood, teaching, and learning*. London: Routledge.

Britzman, D. P., & Pitt, A. J. (1996). Pedagogy and transference: Casting the past of learning into the presence of teaching. *Theory Into Practice, 35*(2), 117–123.

Todd, S. (Ed.). (1997). *Learning desire: Perspectives on pedagogy, culture and the unsaid*. New York, London: Routledge.

CHAPTER 3

DIAMONDS IN A SKULL

Unpacking Pedagogy
with Beginning Teachers

Tony Cotton

**with Corinthia Bell, Lauren Betts, Rachel Cartwright,
Rachael Dean, Amy Howard, Katie Pidgeon,
Joanna Thompson, Laura Willis, Deborah Silberstein,
Hannah Stonehouse, Suzannah West, Jamie Wilcox**

*Many primary teachers still need better knowledge of mathematics if they
are to be enabled both to help the lowest attaining pupils reach the expected standards
and to challenge the highest attainers.*

—Ofsted, 2008, p. 55

*Confidence and dexterity in the classroom are essential prerequisites for the successful
teacher of mathematics and children are perhaps the most acutely sensitive barometer
of any uncertainty on their part. The review believes that this confidence stems
from deep mathematical subject and pedagogical knowledge.*

—Williams, 2008, p. 2

*Learning mathematics is completely different from teaching mathematics.
I love teaching mathematics.*

—Trainee teacher

Unpacking Pedagogy: New Perspectives for Mathematics Classrooms, pages 43–63

ABSTRACT

In the June of 2008, Sir Peter Williams published the outcomes of his investigation into the teaching of mathematics to children for the ages of 3 to 11 in schools in England. This report was commissioned by the Department of Children, Schools and Families (DCSF) to explore the most effective pedagogy for teaching mathematics—effectiveness was to be measured by the progress that children made in their learning. The quotes that open this chapter show a developing consensus from the Office for Standards in Education (Ofsted)—the government inspectorate of schools—and the Department with responsibility for education. This is the view that effective teaching of mathematics is inextricably bound to individual teachers' personal mathematical subject knowledge. During 2008 and 2009, I worked with a group of trainee teachers who had elected to take part in a curriculum development project to support them in developing their teaching of mathematics. All the trainee teachers had expressed a certain "lack of confidence" in their own subject knowledge and hoped that the project would boost their confidence as well as their practice. It is interesting that this group of trainee teachers did not share the view of Ofsted and the DCSF that their own learning of mathematics was directly linked to their teaching of mathematics as they set out on the project.

This national research project was led by The National Centre for Excellence in the Teaching of Mathematics (NCETM), with partners including the Training and Development Agency for Schools, York Saint John University, and Leeds Metropolitan University. The project involved 20 undergraduate students who identified themselves as "lacking in confidence" in personal mathematics subject knowledge. A series of seminars supported these undergraduates in developing their subject knowledge. These undergraduates will also be supported in schools to develop their range of approaches to learning and teaching mathematics. In this sense, the project aims at developing the pedagogical approaches of the trainees through developing their confidence in their own understandings. This may be seen as developing their teaching repertoire within their professional subject knowledge as a result of working at their personal subject knowledge. Using Shulman's (1986) terms, the aim is to develop pedagogical subject knowledge through a focus on content subject knowledge.

This chapter evaluates this process through the lens of identity. The chapter will explore constructions of identities of *mathematics learner* and *mathematics teacher*, focusing on the impact that identity work has on classroom practices. The question may be articulated in a number of ways: How do I develop as a teacher? How does my teaching develop as a result of my working at mathematics? How does my changing view of myself as a mathematician impact my view of myself as a teacher of mathematics and how does this evidence itself in my practice?

Research Process

There has been substantial previous work exploring the mathematical subject knowledge of trainee teachers and teachers at the beginning of their careers (see Goulding et al., 2002; Murphy, 2006; Rowland et al., 2000). This work does not point to any clear conclusion. Indeed, Rowland et al. conclude their study by suggesting that the "precise role of mathematics subject knowledge per se remains unclear" (p. 16), the only suggested link being that teachers who "lack" subject knowledge perform "poorly" in their teaching when assessed at the end of their time in training. Goulding et al. (2002) also identified a link between insecure mathematical subject knowledge and poor teaching. Carol Murphy (2006) suggests, however, that trainee teachers often do not perceive any relevance of audits of their subject knowledge during their time in training.

Tony Brown and his colleagues (Brown et al., 2004) explored issues of identity among trainee teachers as they studied mathematics during their first year at university. They suggested that it is very difficult for trainee teachers to make their own sense of mathematics teaching and learning. Rather, they are "interpellated by multiple discourses" (p. 177), and in trying to identify with recognized constructions of mathematics and mathematics teaching and learning, teacher trainees find it difficult to produce an identity of their own. More recently, I have explored the link between learner identity and mathematics learning (Cotton, 2008). I suggested here that the collaborative exploration of identity construction undertaken in schools, with the learners themselves, supported a developing articulation of personal identity through an emerging language of critique, which, in itself, could begin to connect learners to mathematics. This chapter extends these ideas to work with beginning teachers on analyzing their own constructions of identity, their views of mathematics, their learning of mathematics, and their mathematics teaching.

During the research process, I worked with twelve teacher education students who had put themselves forward to work on the curriculum development project described above. These students were not part of a mathematics "specialist" group—indeed, most of them described themselves as "lacking confidence" in teaching mathematics and hoped to become more confident through their engagement with the project. Initially, I asked the group to explore their identity using a technique often used as a drama warm-up activity. I asked each member of the group to ask me the question, "Who are you?" Each time they asked me this question, I responded with a different facet of who I am, for example, teacher, parent, musician, frustrated football fan, and so on. This offered a model of the complex natures of our identities and brought to the surface the multifaceted nature of our view of ourselves. The students then split into groups of three to carry out and record the activity. We also explored, through small group discussions the students' perceptions around success in mathematics learning. Here, they identified individuals whom they had met and whom they would describe as "good at mathematics." Through this guided discussion, we explored the students' perceptions of what "being good at mathematics" meant.

In two extended group discussion sessions, we developed and explored, in some depth, metaphors for "mathematics" and for "learning mathematics." Individuals in the group drew images or described images in response to the prompts, "Mathematics is like..." and "Learning mathematics is like...." This use of metaphor allowed the participants to draw on personal narratives to explain in detail the metaphors that best supported their images of mathematics and mathematics learning. Finally, the group contributed to the chapter by offering extended vignettes to illustrate the analysis. Through engaging in the writing of this chapter as a collaborative venture and searching for ways in which an authentic student voice can be heard within the chapter, new ways of looking at familiarities are offered for the purpose of change.

INTRODUCING IDEAS

This section explores in detail the underpinning theoretical ideas that will inform the ideas in the practice section. Although seemingly uncontroversial, the area of subject knowledge—in particular, the way a teacher's subject knowledge contributes to the pedagogical choices that are made—is discussed in detail. The impact of individual perceptions of personal subject knowledge and its relationship to personal views of teacher identity are also explored. The first part of this section discusses the particular view of subject knowledge that I am taking in this chapter, and the second part outlines my theoretical approach to teacher identity.

What Do I Mean by Mathematical Subject Knowledge?

An intense interest in teacher subject knowledge and pedagogical practice was initiated by Lee Shulman in 1986, with the introduction of his idea of *pedagogical content knowledge*. Shulman suggested that teachers draw on three forms of knowledge in order to teach effectively. The first is a "deep" knowledge of the subject itself; second, teachers need an understanding of the curriculum they are expected to teach; and third, they must be able to draw on an understanding of the range of pedagogical choices open to them that may support learners in coming to an understanding of the content.

So, in order to explore the way in which "subject knowledge" interacts with teachers' views of themselves as teachers of mathematics, and, in turn, how this impacts the pedagogical choices they make, it is helpful to view subject knowledge as a triplet. This triplet is made up of:

- *Subject Matter Knowledge (SMK)*: This can be described as the extent to which trainee teachers see themselves as having a "deep" understanding of key mathematical concepts. In Richard Skemp's terms (Skemp, 1977), they have a relational understanding of mathematics, rather than an instrumental understanding. That is, they can move beyond a mechanical, or rote view of mathematical processes to see the links and connections between the different areas of mathematics.
- *Curricular Knowledge (CK)*: This can be seen as the extent to which the trainee teachers are able to articulate the demands of the curriculum framework within which they are expected to work. For this group of trainees, this curricular framework is contained within "The Primary Framework for Teaching Literacy and Mathematics," a framework designed and closely monitored by the Department for Children, Schools and Families in schools in England. This curricular knowledge also encompasses cultural expectations built into classroom practices. These expectations become evident when trainee teachers interpret choices they are making as "high risk" or cutting across what they perceive as "classroom norms." Andrew Harris (2006) describes curricular knowledge as "grounded in and constrained by, classroom experience, values, and beliefs" (p. 31).
- *Pedagogical Content Knowledge (PCK)*: For Shulman, PCK can be described as "The most powerful analogies, illustrations, examples, explanations and demonstrations—in a word the ways of representing the subject which makes it comprehensible to others" (Shulman, 1986, p. 9). For the trainees working on this project, PCK would involve the examples they chose to use to introduce a particular concept, the particular pedagogical approach they used, the ways that they grouped their learners, and the way in which they structured their instruction and teaching. In particular, PCK is an awareness of the choices that they made and why they made those particular choices.

The research discussed in this chapter explores the intersections of these forms of subject knowledge and the ways in which the triplet combines and separates to influence the trainees' views of themselves as mathematicians and mathematics teachers, and the impact this has on their teaching. I draw on ideas from Lacan to illustrate these intersections.

Identity

For Lacan (1977), the construction of identity begins with a young child unaware of the social situation into which it has been born. Lacan describes the unconscious process through which the developing child takes on the language of its surroundings through recognizing that there are others on which it depends, but who are outside its control. This "taking on" of language is a means of expressing desire and meeting needs. Lacan describes the *imaginary order* of awareness, which precedes language. During this stage, the child begins the process of identity separation through investing significance in particular events/objects particularly linked to the mother. This stage is interrupted when the child makes an identification with its mirror image. Through the *mirror stage*, the child comes to identify with an image outside itself; this image can be its own mirror image or the image of another. The apparent completeness of this image gives the child mastery over the body. For this group of trainees, there is a range of "significant others" with whom they may identify. In a sense, they use three mirrors in which to "reflect" their pedagogical practice.

Many trainees have clear images of what they describe as their "best teacher." This is a person who inspired them as a learner and whom they wish to emulate in their own careers. They can describe with clarity how this teacher operated. In a sense, this is a vision to live up to. We could see this as an *heroic other*.

There is also an amalgam of practices and beliefs that they have been exposed to, both as a learner in school and as a trainee teacher. These are activities they have observed in school, or been exposed to in the university course that resonate with their understanding of best practice. This set of practices is often embodied by the trainees and described as "good practice"—these are ideas they want to try. We could see this as an *iconic other*.

Finally, there is a set of "standards" against which the trainees know they will be assessed. For this group of trainees, these are the 33 standards laid down by the Training and Development Agency for Schools (TDA, 2008). Much of the training program for the trainees has been mapped against these standards, and feedback on their teaching will have been given in relation to these standards. We could describe this as the *standard other*. It is against these mirrors that the trainee teachers construct their identities.

Morwenna Griffiths (1995) describes "self-identity" as a web. For these teachers, this web is made up of everything they bring to the classroom, together with their sense of how they see themselves reflected in the mirrors I described above. For Griffiths, the construction of identity is "partly under guidance from the self, though not in its control" (p. 93). Individuals only exist within the communities of which they feel they are members, and individuals exist differently within different communities. Identity here is slippery and cannot be pinned down. It can, however, be explored. Tony Brown and his colleagues (Brown et al., 2004) describe identity in a similar way. They suggest that identity is not something we should attribute to individuals; rather, identity is something that people "use to justify, explain and make sense of themselves in relation to other people, and to the contexts in which they operate.... [I]dentity is a form of argument" (p. 167).

In an earlier paper exploring learner identity and mathematics, I suggested that by investigating the process of learners' self-construction of identity in connection with others, processes of learning within the classroom can be exposed (Cotton, 2008). The struggle for meaning takes place between the teacher and the learner over what it means to be a learner. In this chapter, there are two groups of teachers and learners: the trainees and their pupils, and the university tutors and the trainees. In Vygotsky's (1986) terms, the trainees develop a reflective consciousness:

> School instruction induces the generalizing kind of perception and thus plays a decisive role in making the child conscious of his own mental processes. Scientific concepts, with their hierarchical system of interrelation, seem to be the medium within which awareness and mastery first develop, to be transferred later to other concepts and areas of thought. Reflective consciousness comes to the child through the portals of scientific concepts. (p. 171)

Such a reflective consciousness is not only available to the child, it is also available to teachers through an exploration of their own practice. It becomes clear later in the chapter how the trainee teachers are interpolated by multiple discourses—there are tensions and inconsistencies in the ways the trainees position themselves and are positioned by the contexts in which they work and by those with whom they work.

The following section explores these ideas of identity and of subject knowledge through the eyes of the trainee teachers. Much of the text is offered in the words of the trainees—that is why they appear as co-authors of the chapter. If I see the trainees as authors of their identities—or, at the very least, as co-authors in Griffiths' terms—then, clearly, they should be acknowledged as the tellers of the stories that follow. The following section draws on the trainees' stories to exemplify and analyze the theoretical ideas of subject knowledge and identity through their practice.

APPLYING IDEAS

> Narratives help us to represent the world. They also help us to remember and forget both its pleasures and its horror. Narratives structure our dreams, our myths and our visions as much as they are dreamt, mythified and envisioned. They help us share our social reality as much by what they exclude as

what they include. They provide the discursive vehicles for transforming the burden of knowing to the act of telling. Translating experience into a story is perhaps the most fundamental act of human understanding. (McLaren, 1995, p. 92)

Peter McLaren suggests that we structure our world through stories—through the stories we tell about ourselves and how we operate in the world. I would suggest we also structure *ourselves* through stories—our image of ourselves is constructed through the stories we tell ourselves; our image of how others view us is constructed through the stories they tell us about ourselves. Maggie Maclure (2003) supports the view that identity is built through narrative. She sees both self and the social life of our "selves" as constructed through talk. Perhaps, more importantly, she suggests that through "talk," we can both (re)constitute ourselves and exercise agency. Through coming to an understanding of our understanding of our identity, we come to feel as though we can act in the world.

The trainees worked in threes at the beginning of the process to uncover their personal "webs" of identity by responding to the repeated question, "Who are you?" posed by colleagues. This activity can illustrate the many facets to our identities. Half the trainees described themselves as "trainee teachers," although half used a more generic term to describe themselves such as "student" or "Leeds Met student," identifying with the institution. However, the facet of themselves as trainee teacher was not explored further through their descriptions of themselves. Jamie (see Table 3.1) describes himself as a "maths champion"—the project is called "Inspiring Maths Champions," so this offers a direct link to the project. Only one other trainee used the word *mathematician* to describe herself. Interestingly, when Jamie revisited his description of himself through his "I am" statements, he suggested he had tried to project a version of himself he didn't always acknowledge. He said: "I'm not really like that—I think I've tried to project a very male view of myself. I don't really know why." He also wondered if he had projected a particular view of himself as he saw teaching as an "act." "You're always acting," he said, "or at least I am; the minute I start teaching I am in role." This exemplifies the multifaceted nature of Jamie's "authoring" his identity. And the shifting nature of our views of ourselves—he acknowledged that in certain situations, he uses specific signifiers that emphasize his maleness— "I am a Stoke City fan"; "I'm a fantastic son." He wondered if this is related to the small number of males in this course, and to the fact that he is the only male in this particular group.

I would suggest that for many of these trainees—indeed, for many teachers—an identity as "teacher" is only one part of the image of self. When we act as teachers, we negotiate the extent to which the other facets of

TABLE 3.1 Who Am I?

I'm Laura	I'm Jamie
I'm a trainee teacher	I am a man
I love animals.	I'm a Stoke City fan
I lead holiday clubs	I'm a fantastic son
I'm a busy person	I'm a hated brother
I'm a clean freak	I wish I had been born in the late 70s
I'm getting old	I'm a maths champion
I'm brunette	I'm a teacher in the making
I'm a soap fan	I love to travel the world
I'm a spendaholic	I will move to New Zealand (FACT)
I fall in love easily	I'm sarcastic
I'm romantic	I'm moody
I'm good with computers	I'm a jealous person
I'm boring	I'm a student whose time is up
I'm passionate	I'm non academic
I'm awful at sports	
I'm unfit	
I'm friendly	
I'm a big Westlife fan	
I love pink	

identity impact our practice. When teaching mathematics, our sense of self as mathematician also comes into play. For the majority of these trainees, *mathematician* was not a word they would choose to describe themselves. To explore this further, we examined how the trainees felt about themselves as learners of mathematics.

Myself As a Learner of Mathematics

To explore the trainees' view of themselves, I asked them to sketch or to describe a metaphor that helped them describe their feelings about learning mathematics. Jamie drew an image of a skull with a dagger inserted into the mouth (see Figure 3.1). In the eye sockets and littered around the skull are diamonds and other jewels. Jamie used this image to talk about his experience as a learner at school:

> Learning maths was like a dagger going through my skull—no matter how hard I tried, it wasn't good enough. This was really frustrating, 'cause I knew there were really important things to discover, and when I'm teaching and the kids really get it, it's like they've found diamonds. I don't know whether it was my fault or the teachers'.

Figure 3.1 Jamie's illustration.

The image supported Jamie in moving from his image of himself as a learner to himself as teacher. He suggested that he could support children to "pick out the jewels in the skull through questioning"—he also saw "jewels/maths knowledge" as having to be excavated from the past. Jamie agreed that he still viewed mathematics as "scary" and somehow in need of "taming."

After Jamie told this story, the other trainees laid the blame squarely with the teacher. One of his colleagues said, "The teacher should have found different ways to explain it." Another colleague was damning of her teachers, saying, "I was OK until I got into Year 10 (15 years old)—then I had this idiot, and that was the end of me learning maths." All of the trainees except one reported negative feelings about learning mathematics, although these were all related to their experiences as secondary age learners. They found it very difficult to describe memories of learning mathematics specifically in primary school. They could describe positive and negative memories of themselves as learners, but not specifically as learners of mathematics. They did, however, draw on their negative memories of teaching and learning in secondary school to take responsibility for providing positive experiences for the children with whom they are working.

Another image that the trainees identified with was one of an Escher staircase with a range of "emoticons" placed around the staircase (Figure 3.2). The size of the emoticon related to the importance or prevalence of that particular emotion. For Corinthia, learning mathematics was like "a maze with lots of stairs and me near the bottom." However, she viewed this as a challenge, rather than something to be feared.

Corinthia was one of two trainees who had found learning mathematics a positive experience. She remembered feeling slightly uncomfortable about this at school. "People think you're going to be a brain surgeon or

Figure 3.2 Escher with emoticons.

something if they think you're good at maths." She reported that she was happiest when left alone to try "and figure things out." She had felt particularly uncomfortable in an activity that they had been asked to work on as a group in a recent training session. She said, "My problem is being put on the spot—I can't think then, lots of little warnings run through my head the moment I'm asked a question. I was literally running for my life during that activity." This comment drew agreement from the group. One of the aims of the training sessions they had attended was to support them in both seeing learning mathematics as a communal activity and learning to use activities in their teaching that would encourage learners to work together. They had found these activities threatening, even though they were supportive of each other. They could see the value of the activities, but they cut across their previous experience, and the feelings of insecurity in their own knowledge that the group brought with them from their own learning made it difficult for them to engage with the activity openly. It is likely that there was a tension emerging here between the view of *standard other*, which many of the group saw being played out in the classrooms they were placed in, exemplified by an individualistic approach to learning mathematics, and a positivist view of mathematical knowledge, and a new signifier being presented for the *iconic other*. The trainees were experiencing a symbolic process of identification—lacking the (however fantasized) "whole-image" that a mirror might portray.

The next section describes the metaphors the trainees used to describe their image of themselves as teachers of mathematics.

What is it Like to Teach Mathematics?

I asked the trainees to respond with a metaphor to the question, "What is it like to teach mathematics?" One trainee offered this image:

> Teaching a hard mathematical concept is like watching your team win a cup final in the 90th minute. It can be frustrating for long periods, nerve wracking, and you feel the intense pressure. However, when the children finally understand, you realize, as a teacher, why you tried so hard.

This led the group into a discussion of the mathematics curriculum they had experienced—an example of a move between personal pedagogical content knowledge (as experience) and curriculum knowledge (as experienced). There was agreement that they could not see the relevance of much of the mathematics curriculum that they encountered as learners, and they also suggested that they found it difficult to make the curriculum they were teaching engaging and motivating for their learners, although they did see that it was important to do so. One of the group remembered one of the teachers she was working with during a placement saying to one of her pupils, "'Why do we have to do this?' It's because it's in the Framework— that's why you have to do it." There was agreement around the group on this: Mathematics is something that is done to you, or that we do to other people. Similarly, the mathematics curriculum described here is something that is imposed from an outside source—and one of a teacher's tasks is to make it as interesting or as relevant as possible, despite a lack of inherent relevance or interest. This feeling of powerlessness over the content of the curriculum emerged later. The trainees also saw an inconsistency between their images of themselves as "teacher"—"I want to get the kids excited about mathematics and to see how important maths is in their real lives"— and the content that they are required to teach.

Corinthia offered a metaphor for what it is like to teach mathematics, once again using emoticons (Figure 3.3).

She described teaching mathematics as striving for the pot of gold at the end of the rainbow—the "pot of gold" being the point at which she knew the pupils had "got it." For Corinthia, this is represented by the light bulb illuminating—a moment of triumph. The range of emotions along

Figure 3.3   Corinthia's illustration.

the rainbow represents the phases that Corinthia goes through in preparing for her teaching. Her starting point on the rainbow is dependent on how confident she feels with her own subject knowledge. In turn, her own confidence impacts the pedagogical choices she might make. She said, "If I'm doing something that I don't know, if I'm out of my comfort zone, I stick to my plan."

The contrast between mathematics and other subjects was drawn out later in the discussion. I questioned the "relevance" of teaching History as a way of exploring the trainees' perception that mathematics was difficult to engage with, both as a teacher and a learner, because it was not relevant to lived experiences. They responded, "You need to know history because it makes you a more rounded person, something like understanding algebra doesn't make you more rounded." This group of trainees had not experienced learning mathematics as something that was interesting for its own sake, or did not feel better equipped to interpret and understand the world or themselves through their mathematics learning, and so found it difficult to see this as a responsibility of a teacher.

Throughout the discussion, the rewards in teaching mathematics were clear; the metaphor of "seeing a light come on" was used repeatedly. One trainee talked about the difference between teaching Mathematics and English. She said, "I think you see the light going on teaching maths; when you're teaching reading, you don't get light bulbs going on." This sort of instant feedback was a source of pleasure and reward. Another trainee described the moment she decided to want to teach: "I clearly remember actually teaching a child how to do multiplication—that's what made me want to teach." This takes us back to the initial metaphor—mathematics teaching as striving for a moment of success, the discovery of a diamond in a skull, or the scoring of a goal in a cup final.

For the majority of these trainees, mathematics was seen in terms of struggle—but a struggle for a worthwhile reward. There is a reward for the learner in suddenly realizing that you have come to an understanding of something new. There is a reward for the teacher in seeing a learner realize they have come to understanding. Discussion also centered on their relationships with the children that they taught. Here, the struggle was their own prioritizing of relationships with the children they were teaching and a curriculum, in the sense of mathematical content, which they could characterize as "relevant" to their learners. There was a disjuncture for them here: *Relevance* was seen as a key signifier for the appropriateness of curriculum, but this group of trainees struggled to attach this signifier to the mathematical ideas they were exploring with their learners.

The following section follows two of the trainees into their classrooms to further explore how their images of themselves as learners and teachers, and how their image of mathematics as a subject, impacts their planning and teaching.

Myself as a Teacher of Mathematics in Practice

> *It's very hard because, basically, you just have to be like the class teacher—I mean, that's the way that you're going to get a good grade at the end of the day.*
>
> —Laura

I worked closely with two of the trainees, Rachel and Rachael, as they moved into the classroom. This allowed me to begin to get a sense of how their views of themselves as teachers of mathematics impacted their practice. It also offered a reminder of the constraints and pressures that exist as we move from theoretical views of pedagogy to negotiating a way to insert ourselves as teachers into classrooms. In many senses, classrooms are "owned" by others, whether they be the class teacher, the head of the school, the parents, or the learners that have been operating within the classroom before our arrival.

Before observing Rachel and Rachael in their classrooms, we spent some time talking about the process of planning for the lesson I was about to observe, and reflecting on how they were negotiating the relationships with their class teacher to engage in activities that did not follow the usual pattern for the classes they were teaching. As they worked with the learners, I used an observation sheet, which I had subdivided to focus my attention on the ways in which the themes I introduced earlier came to the fore during the lesson. I tried to notice evidence of SMK, PCK, and CK in the classroom practice. I also tried to see if the facets of identity, *heroic other, iconic other,* and *standard other* were evident in action.

One of the tasks I was set for this chapter by the editor was to support readers in advancing their understanding of mathematics pedagogy. As you engage with this section, I would ask you to use the text as a reflective tool to ask yourself questions around your own practice. Perhaps I am offering a way to advance your understanding of your own pedagogical practice? Observing Rachel and Rachael certainly helped me advance my understanding of how I work with learners and why I might make the choices I do. I attempt to capture a sense of that in the next few paragraphs.

Curricular Knowledge in Practice: Earlier in the chapter, I described the tension that the trainees felt between the content of the curriculum that they were being asked to teach and the ways of working that they had been introduced to in the workshops. It proved to be problematic for Rachael and Rachel to negotiate their way through the intersection between *iconic other* and *standard other* in an assessed school placement.

I asked how they had decided which area of mathematics they were going to focus on. Rachel explained that the class teacher followed the framework that was laid down by government, as interpreted by a commercially produced program. This program took each of the objectives outlined by the framework and offered a series of lessons which would "cover" these objectives. Rachel would look at the series of lessons in the program and utilize the activities that met her view of appropriate teaching and learning: "I like to make sure the children are active and can see the relevance to their everyday life." If she deemed the activities inappropriate, she would plan different activities. At no stage did she question the appropriateness of a piece of curriculum. Similarly, Rachael had been asked by her class teacher to teach "something to do with measurement." To support herself in deciding what would be appropriate, she drew on the government framework to help her focus her planning. In many senses, there is no alternative to this. As all the learners' mathematical experience prior to the lesson had been defined by the framework, the learners' prior experience is prescribed. Teachers also feel a responsibility to make sure that colleagues who work with young learners once they move on out of their classrooms can be secure that an appropriate curriculum has been followed. So, for Rachel and Rachael, the curriculum is a given, and is an imposed given. It is not a constraint they can work outside. The mathematics that we teach is externally decided. As with all constraints, this view of curriculum bounds our thinking.

A further example of the trainees' bounded view of an externally imposed curriculum came when they attended the annual conference of the Association of Teachers of Mathematics. Several of the trainees presented their work at this conference. The opening keynote explored knot theory, in an active and involving way. There was much discussion around the use of modulo arithmetic as a notation to explore knot theory. Modulo arithmetic does not appear on the UK syllabus, and so was a new area for the group. The group found it difficult to see modulo arithmetic as relevant or interesting—for them, it was only of interest to "maths geeks." It was certainly not something they would wish to explore any further.

I found myself wondering what the possibilities are for not viewing the curriculum as given. Indeed, much of my energy as a teacher educator is taken up by finding effective ways of teaching the curriculum we are given and taking it further, rather than questioning the nature of that curriculum.

Subject Matter Knowledge in Practice: Rachael is one of the two members of the group who described herself as confident in her subject matter knowledge. (Corinthia is the other). Rachel, on the other hand, had elected to come on the Inspiring Maths Champions program to help develop her confidence in her own knowledge. There were some differences in the ways in which their views of "doing mathematics" drew on this experience. Rachael was very precise in her definitions—she confidently and consistently used technical vocabulary to model accuracy and preciseness. Toward the end of the session, she asked the children to justify their answers. They were exploring the areas of rectangles with a constant perimeter, and had realized that a square gave the greatest area. Rachael asked for a "proof" for this conjecture, and was prepared to leave this hanging as a question they could explore later. However, one of the young learners—she was working with 9/10-year-olds—noticed a pattern. He saw that if you started with the rectangle with width 1 cm, to change it to a rectangle with width 2 cm, you "take 1 square and move it, but then you have to fill up the whole row, so that adds lots more squares." This description convinced his classmates that there was a good reason why a square gave the largest area. This clearly delighted Rachael. Her view that proof and justification is important had been supported by the underlying philosophy of the Inspiring Maths Champions project, which had provided her with the activity and the methodology for the lesson.

Rachel was also exploring an activity that had been introduced to her by the Inspiring Maths Champions program. She had given the children a number—each had a different number card with a numeral between 1 and 20 on it. When Rachel described a number property, any child with a number with that property had to swap places with another child with a number with the same property. Rachel's focus was on the process of the activity; she explained that she had tried out the activity initially and it "had been a disaster," however, she had persevered, adapted the process, and the activity supported the children in asking thoughtful questions. Rachel arranged her group in a circle and asked all the children who had a card with a multiple of 3 on it to swap places. I overheard one boy say,

"I'll never be a multiple—cause I'm a prime number." She asked those who were "a multiple of 2 plus 3" to swap. Another boy said, "That will just be odd numbers except 1." And finally, for "a multiple of 5 less 1," a girl wondered if this was "the same as the 4 times table." Here, the activity, through the questions that Rachel was asking, was supporting the children to generalize. Although Rachel's focus on the pragmatics of organization didn't allow her to hear these comments in real time, she realized that these were a measure of the effectiveness of her teaching in discussion after the lesson. She also realized that her developing subject matter knowledge was supporting her in planning "good" questions that would challenge and extend the children's thinking.

There is some evidence here to suggest that by engaging these trainees in working at mathematics and in actively questioning and reflecting on their learning, they have been able to change the nature of their own subject knowledge, in terms of their beliefs about what it is to be a mathematician. They were introducing their learners to a form of mathematics that they had recently experienced, themselves, as learners. This allowed them to be both tentative, in terms of the possible outcomes, while being secure in the process. They knew, or they trusted, that interesting things would happen. They are also moving towards a belief that mathematics is about questioning, exploring, and justification, and so these were elements that they inserted into their classrooms.

Pedagogical Content Knowledge in Practice: A colleague of mine once suggested that "the only thing we don't seem to learn from experience is how to learn from experience." It seems to me that the only way we can decide which are *the* most powerful analogies, illustrations, examples, explanations, and demonstrations is by working with a range of analogies, examples, explanations, and demonstrations, and reflecting on how effective they have been in supporting learners to develop their mathematical understandings. Rachel had a clear view as to the type of activities that she wanted to work with: "I like to use real-life situations, and I think it is really important to get the children actively learning." She also used techniques she had observed and which seemed to her to be effective—children modeling their thinking processes with others through taking on a pseudo-teacher role at the front of the class; working with empty number lines; drawing on peer support to work at misconceptions. Rachael, on the other hand, had made decisions about the ways she would organize the groups and used formative assessment throughout the session to gauge her learners' understandings. However, both were still exploring these strategies. As beginning teachers, they did not yet have a wide experience to draw on in order to decide if these strategies would offer the most effective routes into specific mathematical areas. Their choices were based mostly on what I described earlier as an *iconic other*—these were activities and techniques they had been taught, either by tutors at University or through the Inspiring Maths Champions Project.

It appeared to me, however, that Rachel and Rachael were making clear pedagogical choices. They were both prioritizing ways of working that allowed them to develop caring and respectful relationships with their learners. As new teachers, with groups of children they had only recently met, their first priority was to develop and sustain these relationships. An understanding that backgrounded any belief in ways of teaching mathematics was that effective teachers support their learners emotionally and socially.

A question raised here for me deals with the extent to which I ask trainee teachers I work with, or, indeed, mathematics teachers I work with, to reflect on the pedagogical choices we make. Am I supporting people to learn how to learn from experience? Similarly, if we simply focus on developing mathematics pedagogy, are we at risk of backgrounding pedagogies with a focus on developing relationships between teacher and learner?

Teacher Identity in Practice. In the previous paragraph, I suggested that the *iconic other* was present in the choices that Rachel and Rachael made. They drew on ideas and strategies that had been presented to them as good practice, even if these were not viewed problematically. This *iconic other* had subsumed the *heroic other*, in that neither described teachers they had worked with as models for their current practice. It seemed that the "best teacher" that may have inspired them to become trainee teachers had faded into the background during the four years of the course. However, the overwhelming issue for Rachael and Rachel, in terms of their identity as a teacher, was the *standard other*. Many of the trainees described how they felt that they had to take on the mantle of the class teacher in order to be successful. This led to tensions, as sometimes they "could not be themselves." A bigger tension was described by Laura when she suggested, "I don't really know what sort of a teacher I really want to be yet. I suppose I will be able to be myself when I have my own class." Rachael described this tension in another way, saying, "I guess you just go with the style of the teacher you are with. I will move into my own style next year."

LAST WORDS—FOR NOW: UNPACKING PEDAGOGY AND SUBJECT KNOWLEDGE

When we reflect on the theoretical models offered, a number of issues begin to surface. First, the Shulman triad raises some intriguing questions. How might we work with the idea of curriculum knowledge when, for many teachers, the curriculum is a given? The choices that these trainees were making were about *how* to teach, not *what* to teach. There was never a sense of challenging the appropriateness of the curriculum—indeed, in a UK classroom with a focus on outcomes and high stakes testing, it could be seen as irresponsible to make any deviation from the given curriculum. The notion of subject matter knowledge had allowed me to notice ways in which the trainees had refined their view of what being "mathematical" might be and ways in which they could support their learners in becoming mathematical, in this sense.

Finally, I would suggest that Shulman's notion of pedagogical knowledge is problematic, if viewed as something which is fixed or can be attained. We cannot find "the most powerful" ways of working with children. A better aim might be to find ways to give ourselves space to explore how effective our ways of working have been—and against what criteria we are measuring effectiveness. And for me, most importantly, Shulman's triad does not allow us to focus on developing caring and respectful relationships with our learners. Perhaps the idea of subject knowledge needs re-theorizing to bring this to the fore.

I am moving towards an alternative approach to "unpacking pedagogy," which draws on theorizations from the coaching of athletes. A triad is still offered, but it is a triad with a different key focus. Initially, the goal of the coaching process is negotiated and agreed. Coach and athlete come to a shared view of "success"—in order to support the athlete in achieving success, the coach draws on three areas of knowledge:

- A conceptual understanding of the student and the student's needs
- A conceptual understanding of the subject matter
- A conceptual understanding of teaching and learning environments

I would suggest that the latter two areas coincide with Shulman's SMK and PCK, with CK replaced by a focus on the learner and their specific needs—perhaps the focus on care that the trainees found lacking in the models of pedagogical knowledge offered to them.

Second, I offered three ways of looking at developing teacher identity, using the Lacanian idea of "taking on" language and actions. I explored the idea of *heroic* and *iconic others*. My work with this group of beginning teachers would suggest that these two "beings" merge into one—it may be that our beliefs, our sense of self emerges as we work with teachers we trust, but these individuals fade away as we develop our own ways of working. Finally, for these beginning teachers, their own emerging identity as a teacher and a teacher of mathematics is currently held at bay as they strive to complete their time as trainee teachers. They feel the need to "take on" the language and actions of their class teachers so that they can achieve the qualification of "Qualified Teacher Status." Only when they have a class of their own will they be able to "develop their own style." As Saramago (2002) suggests:

The journey is never over. Only travellers come to an end. But even they can prolong their voyage in their memories, in recollections, in stories. When the traveller sat in the sand and declared: "There's nothing more to see," he knew it wasn't true. The end of one journey is simply the start of another. You have to see what you missed the first time, see again what you already saw, see in springtime what you saw in summer, in daylight what you saw at night, see the sun shining where you saw the rain falling, see the crops growing, the fruit ripen, the stone which has moved, the shadow that was not there before. You have to go back to the footsteps already taken, to go over them again or add fresh ones alongside them. You have to start the journey anew. Always. The traveller sets out once more. (p. 443)

I hope to start a new journey with this talented group of teachers. Perhaps together we can explore the ways in which they develop their identities as teachers of children and as teachers of mathematics as they start their careers.

ACKNOWLEDGMENT

I wish to acknowledge all the trainee teachers who worked on this project with me as co-authors. The chapter was put together through a series of workshops and interviews, as well as the observations, and the group has made many suggestions as to the form the text should take. As well as acknowledging the joint ownership of these ideas, I hope this also allows you to read the chapter through the eyes of these individuals, rather than viewing it as a research report. I also want to thank and acknowledge Jane Barber and Tansy Hardy for their comments on an initial draft of the chapter. I hope I have done justice to their suggestions. Thanks also to Andy Abrahams for a discussion around creating a coaching environment, which helped me strengthen my argument.

REFERENCES

Brown, T., Jones, L., & Bibby, B. (2004). Identifying with mathematics in initial teacher training. In M. Walshaw (Ed.), *Mathematics education within the Postmodern* (pp. 161–181). Greenwich, CT: Information Age.

Cotton, T. (2008). 'What is it really like?' Developing the use of participant voice in mathematics education research. In T. Brown (Ed.), *The psychology of mathematics education: A psychoanalytic displacement* (pp. 183–197). Rotterdam: Sense Publishers.

Goulding, M., Rowland, T., & Barber, B. (2002). Does it matter? Primary teacher trainees' subject knowledge in mathematics. *British Educational Research Journal, 28*(5), 689–705.

Griffiths, M. (1995). *Feminisms and the self: The web of identity*. London: Routledge.

Harris, A. (2006). Observing subject knowledge in action: Characteristics of lesson observation feedback given to trainees. *Proceedings of the British Society for Research Learning in Mathematics, 26*(2), 31–36.

Lacan, J. (1977). *The four fundamental concepts of psycho-analysis*. London: The Hogarth Press.

Maclure, M. (2003). *Discourse in educational and social research*. Buckingham: Open University Press.

McLaren, P. (1995). *Critical pedagogy and predatory culture*. London: Routledge.

Murphy, C. (2006). Why do we have to do this? Primary trainee teachers' views of a subject knowledge audit in mathematics. *British Educational Research Journal, 28*(5), 227–251.

Office for Standards in Education (Ofsted) (2008). *Mathematics: Understanding the score*. London: DCSF.

Rowland, T., Martyn, S., Barber, P., & Heal, C. (2000). Primary teacher trainees' mathematics subject knowledge and classroom performance. *Research in Mathematics Education, 2*, 3–18. London: British Society for Learning in Mathematics.

Saramago, J. (2002). *Journey to Portugal*. London: The Harvill Press.

Shulman, L. (1986). Those who understand: Knowledge growth in teaching. *Educational Researcher, 15*(2), 4–14.

Skemp, R. (1977). Relational understanding and instrumental understanding. *Mathematics Teaching 77*, 20–26. Derby: Association of Teachers of Mathematics.

Training and Development Agency for Schools (TDA) (2008). *Professional Standards for Teachers*. Accessed October 15, 2008 from http://www.tda.gov.uk/upload/resources/pdf/s/standards_a4.pdf

Williams, P. (2008). *Independent review of Mathematics Teaching in Early Years Settings and Primary Schools*. London: DCSF.

Vygotsky, L. (1986). *Thought and language*. Cambridge: Massachusetts Institute of Technology.

FURTHER READING

Brown, T. (Ed.). (2008). *The psychology of mathematics education: A psychoanalytic displacement*. Rotterdam: Sense Publishers.

Maclure, M. (2003). *Discourse in educational and social research*. Buckingham: Open University Press

Walshaw, M. (Ed.). (2004). *Mathematics education within the postmodern*. Greenwich, CT: Information Age Publishing.

CHAPTER 4

THE GOOD MATHEMATICS TEACHER

Standardized Mathematics Tests, Teacher Identity, and Pedagogy

Fiona Walls

ABSTRACT

Compulsory standardized testing regimes in mathematics education are becoming increasingly widespread as systems of monitoring performance of
state funded schools. Such tests are used not only to measure an individual
child's attainment, which can be reported to the clients (parents), but also to
measure the proficiency of teachers and schools—the competing providers.
It is within these economy-driven policy models of education that new identities of "the good mathematics teacher" are emerging. Investigation of links
between standardized testing and teacher identity is important for our understanding of implications for classroom practice. Research involving mathematics teacher identity seeks to clarify the complex and shifting demands of
teachers as individuals acting within the social settings of classroom, school,
community, and beyond. This chapter focuses on perceptions of the "good
mathematics teacher," the part that standardized test regimes play in teacher

Unpacking Pedagogy: New Perspectives for Mathematics Classrooms, pages 65–83
Copyright © 2010 by Information Age Publishing
All rights of reproduction in any form reserved.

identification, how teacher identity is reflected in classroom practice, and how teacher identity is implicated in pedagogical change.

INTRODUCING IDEAS

Identity can be examined and explained using a range of theoretical perspectives, including psychological, postmodern, and sociocultural. Identity theories are argued according to beliefs about where identity is located and how identity is shaped. Côte and Levine (2002) saw this as a *structure–agency debate*, which argues the extent to which an individual's behavior is determined by social structure and how much control individuals exercise independently of such structure. From a cultural studies perspective, Holland, Lachicotte, Skinner, and Cain (1998) viewed identity as a *self-in-practice* arising within the nexus of sociocultural worlds and the world of the individual. They distinguished between *figurative* and *positional* identity, the former described as generic, desired, and imagined, and the latter more specific, located, and relational.

The discussion in this chapter engages the principles of all these theoretical approaches. It is based upon the following theoretical ideas:

- teacher identity engages both the psychic and social
- teacher identity is discursively produced, that is, enacted by individuals within specific social practice
- teacher identity is implicated in pedagogical practice

Key Idea 1: Teacher Identity Engages Both the Psychic and Social

As is every member of human societies, a teacher of mathematics is engaged in a continuous process of identification, both as an individual and as a social professional being. Teacher identity is expressed through teachers' self-explanations, decisions, and actions in their everyday classroom teaching of mathematics. These actions do not take place in isolation; they are molded and delimited by the beliefs and expectations of children, school, parents, community, and state policy. Miller Marsh (2002) describes this process as follows: "[T]he various discourses that define what it means to be a particular type of student or teacher in this particular moment . . . are rooted in the social, cultural, historical, and political contexts in which schools are situated. . . . These discourses of schooling shape what and who schools, teachers, children, and families can become" (p. 460). Where many postmodern theorists perceive identity, such as "self as mathematics teacher," to be produced in social practice, psychoanalytic theories of subjectivity, such as those of Lacan (1977, 2002), have been more concerned with explaining the work of the subconscious, suggesting that it is in the interplay of what Lacan termed the *Symbolic, Imaginary,* and *Real* psychic registers that identity is poised uncertainly as something always in the making—something formed and forming in its own seeking of itself. Lacan pointed particularly to the desire of the self to be "present" as a secure identity. He used the concept of the *Big Other* to describe our consciousness of an omnipresent, ever-presiding entity in our identification. He saw identification as layered like a quilt with what he describes as *points de capiton,* those elements of our lives that serve to catch the layers together in particular settings. In this view, when teachers are confronted with competing versions of what it means to be a good teacher, they engage in both an individual and collective process of reconciliation of currently held beliefs about what it means to be a good teacher and how the self is implicated in these beliefs. Such dissonance compels teachers to contemplate which view best enhances the security of their identification as teachers, allowing them to imagine themselves as successful and effective. New visions of best practice move teachers to reassess their currently held views of the good teacher, since identity exists as an ever-precarious and contestable reality. Lacan draws our attention to how identity, much like a Klein bottle or a Möbius band, is at once both external and internal, in fluid and continuous interplay, since we have no subconscious without a conscious, no *real, imagined,* or *symbolic* without each of the others, existing only in relational interactions in ongoing processes of co-construction.

Key Idea 2: Teacher Identity is Discursively Produced

For postmodern theorists such as Foucault, an individual is not and never can be a single "identity." Through his work, Foucault urged us to *refuse to be who we are,* meaning that although we may be subjectified—created as a subject—by those around us, and by the institutions and signifying social practices in which we find ourselves embedded, we can actively work to accept, to remake, or to reject those constructions of our "selves" that are produced in the processes of discursive engagement. His declaration, "Do not ask me who I am, and do not ask me to remain the same" (Foucault, 2002, p. 19), illustrates that, for Foucault, our view of what it means to be human and what it is to be a "self" is not natural or inevitable; rather, it is a social construct with historical roots. Discourse lies at the heart of Foucault's understanding of how such construction occurs. In his view, schools and schooling belong to an episteme of institutionalization in which discursive games of truth are played, determining what may be said and what is unsayable, how one may or may not behave, and whom one is permitted to be. He argued that we produce ourselves and each other through the continuous and contingent discursive processes that operate within such intersecting social networks as family, workplace, institutions, community, and the political domain. For example, as teachers, whenever we read a novel, teachers' manual, or newspaper item about teaching, watch a film or documentary about schooling, or even participate in a discussion about a classroom incident in the school staffroom, we are engaged, consciously or not, in a discursive process of self-construction as we bring our selves and each other into being in the act of our seeing or hearing. Thus, identity is not something one *has*; rather, it is something one *does.* The view of self as constituted within discourse draws our attention to how this works when we identify and are identified within visions of schooling that produce us as teachers and learners. This process is necessarily social, since nothing we do or say as an individual comes about as a socially disconnected act of the will of the "self." Thus, we are caught in our own realities of how we are produced as teachers and learners within the institution of school, a process which neither crystallizes nor reveals for us a "true" identity. Indeed, a postmodern view holds that there is no such thing. This view of identity as situational and enacted suggests that a teacher may experience differing and/or conflicted views of the *self as a teacher* through discursive mediation, such as discussions with friends and colleagues, classroom interactions with children and parents, and engagement with policy in practice.

Key Idea 3: Teacher Identity is Implicated in Classroom Practice

Researchers interested in identity as formed in the practice of teaching mathematics include Schifter (1996), who explored the multifaceted nature of teacher identity expressed as individual professional narratives constructed in practice. Wenger (1998), and Boaler, Wiliam, and Zevenbergen (2000) focused on the situated and social nature of teacher identity as something built within communities of practice, while Hogden and Johnson (2004) and Van Zoest and Bohl (2005) investigated how specific teacher knowledge, enactment in social situations, and cognitive engagement are implicated in mathematics teacher identity. These researchers saw identity as emergent within specific and specialized arenas of practice, which both produce and are produced by teacher identity. Mendick (2006), who explored the gendered nature of mathematical identification in upper secondary schooling, argues that "the word identity sounds too certain and singular, as if it already exists rather than being in a process of formation" (p. 23). Her preference is to speak of *identity work* or *identification* (from Hall, 1991). These terms capture the nuanced, mutable, and lived nature of identity as situated, as in constant process, as both psychic and relational, and as represented in narrative. In her analysis of student subject choice, Mendick suggested that "'identity work' positions our choices as producing us, rather than being produced by us" (p. 23). In this view, teachers are psychically active and distinctive "selves," as well as socially interactive and connected beings in communities of practice. Teachers choose to act in particular ways, not because of *who* they *are*, but in a continuous *process* of *becoming*. Identity work occurs as teachers choose to engage in particular ways within mathematical communities of practice, thus producing themselves within wider versions of "good" teaching. Viewing identity as "work" is particularly useful for explaining the conflicts that teachers face in explaining themselves, making decisions, and taking particular actions as "good teachers."

APPLYING IDEAS

I now turn my attention to investigating how these theories can be used to help us better understand the processes at work in mathematics teacher identity work in an educational climate of increasing standardized testing. In the following section of this chapter, I bring the key ideas together, since there is much overlap between them and each lends strength to the others.

If teacher identity is, as Foucault suggested, a process embedded in discourse, we might begin by looking at an example of the discursive construction of a good teacher taken from the Editorial in the UK newspaper, *The Guardian* (19 July, 2008):

> Education is at the heart of progressive government. There is a danger though in its very prominence, for the more important it is to the government, the more the government will want proof that its policies work. As a result, tests intended quite sensibly to measure progress have become a curse, a stress for the children that sit them, a much-hated constraint on the teachers that teach them and, when processed into league tables, a controversial way of informing parents about how good schools are. (p. 36)

Although not explicitly drawn, a constructed vision of the good teacher hovers beneath the surface of this statement about teaching and standardized test results. She is centered at the *heart* of the work of *progressive* government. In other words, her work is valuable and inherently worthwhile, yet *endangered* because of control from above, which may escalate into something that is beyond what is reasonable. She is presented as one who *sensibly measures progress*, who is capable of teaching children well without a despised and interfering *constraint* from above, and who sees testing as inflicting a *curse*-like *stress* upon the children in her care. If her school is not perceived to be *good* according to the test results, it does not necessarily follow that she is not a good teacher, herself, since this method of information gathering and dissemination is *controversial*—in other words, contestable.

As readers of such statements, we may find ourselves resonating in varying ways with the vision of the good teacher that this statement calls into being for us. We may find ourselves agreeing or disagreeing with the comments, and, consciously or not, we might perceive ourselves as identities in a process of self-positioning with reference to what we are reading. The key point for us is that tests and league tables are bound up with who we are—indeed, who we are allowed to be—as teachers.

A news item from a later edition of *The Guardian* (19 September, 2008) focused more specifically on the good *mathematics* teacher in a climate of high-stakes assessment:

> Nearly half of all maths lessons are not good enough, according to an Ofsted [Office for Standards in Education] report out today which suggests that even though more pupils are getting qualifications, uninspiring teaching means they often lack understanding of mathematical concepts. . . . Children are being drilled to pass exams and enrol for booster and revision classes, but are not equipped with the mathematical skills or knowledge needed for their future, [the report] will say. (p. 23)

Again we see how this kind of discourse communicates a vision of the good teacher. She is presented as one whose teaching is *inspired* (as opposed to *uninspired*). Good teachers do not *drill* children to pass exams, rather, they *equip* them with *mathematical skills* and *knowledge* which they will *need in the future.* The report is as much about constructing good and bad teachers of mathematics as it is about the tests and exams that bring their bad teaching either to light, or into being, as the case may be.

Public perceptions about what constitutes good and bad teachers, and, in turn, good and bad education policy, are thus discursively produced through many channels, including the media.

In the following section, I look more specifically at how teachers involved in compulsory standardized statewide mathematics assessment talked about their experiences and how their statements can be seen as productive, that is, as teacher identity work, both psychic and social, discursively produced within communities of practice. I look at what implications they signal for mathematics pedagogy. The analysis is based on data collected during a research project that began in 2005 and was extended in 2007. The administrators of a large suburban school in Queensland, Australia were aware of challenges the Queensland Studies Authority (QSA) Numeracy Tests presented for teachers, students, and parents, and requested research assistance in gathering data and compiling a report to present to the QSA, believing a community-based, bottom–up approach to ongoing development in education to be essential. Four Year 5 teachers and five Year 7 teachers were involved in the study, in which they shared their views and experiences of the Numeracy Tests. Shortly after the test had been administered, the teachers discussed the test in small focus groups. Their conversations were recorded and transcribed. The thoughts expressed and issues raised by the teachers during these conversations were analyzed for the ways in which standardized mathematics assessment might be implicated in their identification as teachers of mathematics. Allie, Dan, Col, Karen, Kate, Lee, and Yolanda had been teaching Years 5 and 7 students for many years and were familiar with the tests, but for Jemma and Ralph, this was a new experience. In contemplating their experiences of the test, the teachers were constructing truths about the good teacher of mathematics against which they reflected, refined, and maintained their own identities.

One of the first issues the teachers raised in their conversations was the preparation of the children for the test. They spoke about preparation as something they believed they must do, but something that was problematic at the same time, since the content of the test varied from year to year and the multiple-choice format of the test was something they said was not used in everyday classroom practice. Test preparation, therefore, created challenges and conflicts for the teachers as they juggled competing directives, expectations, and modes of working:

Karen (Year 7): You have to prepare them, but you don't want to scare them. You try to prepare the kids for what's in the test. As you prepare, you think, that in the past there has been a big measurement component, and operations with the calculator. And this year what is the emphasis?— spatial knowledge! Like, it's, "We're testing only the boys this year because you want to see how they're going on spatial knowledge." For us to prepare, because of where [the test] comes in the year, we haven't covered all the concepts so you try to pick up what possibly will be there and spatial knowledge [in the past] has only been [a] fairly small [proportion of the test].

Jemma (Year 7): This was my first time doing the test. I didn't really know how to prepare them. In the few weeks before, you do a lot of practice tests. I worked with number more than anything presuming that this is such a large part of it.

Ralph (Year 7): You want your kids to do the best that they can and you try to help them out and you do your preparation.... It's good to do them in the multiple choice format that they have, just to familiarize the kids. So, I felt that they were prepared, but I didn't think I over-killed them; I was comfortable enough that they had prepared well enough for it.

Col (Year 5): We talk about pressure on the kids, but also on the teachers, as well, because if we don't, you know. . . . We've got to prepare for this, because you're not giving the kids the opportunity to show their best.

Dan (Year 5): You've just got to make sure that you've covered everything, so the kids aren't surprised. . . . It's a stress to make sure that you've covered everything, and you've covered it well enough. . . . We did a lot of practice, especially the numeracy one. I found a really good book and it had quite a bit of multiplication, colouring in bubbles. I felt that it really helped them.

In these statements, the teachers demonstrated the discursive nature of identity work, with the ideal teacher made visible in their contemplation of pedagogical performance surrounding the test. Particular views of this ideal were produced in the telling. As they described and rationalized the decisions made and actions taken, the teachers were, at the same time, talking of themselves, contemplating what it is that constitutes best practice within the demands of the test. They spoke of themselves as those responsible for grooming children for such tests, ensuring children were *familiarized* with the test format without *over-killing* or *scaring* them, their actions described as the seeking of a fine balance between under- and over-preparation, and making sure, at the same time, that the preparation was of an appropriate nature. Seeing themselves in a preparatory role, the teachers talked of how they attempted to anticipate the content of the test, based on their knowledge of previous tests, and to *cover* all the mathematical concepts that they thought likely to be included. Because the test was not presented in a format familiar to the children, the teachers told of how they geared their pedagogical approaches to include several weeks of intense training prior to the test, a practice described as *helping* the children to *show their best.* Failure to train the children correctly was thus constructed in the teachers' discourse as a form of negligence or dereliction of duty.

The teachers talked of the stress that preparation for the test produced, both for the children and themselves. Their talk suggested that they viewed their attempts to prepare the children as being thwarted by the unseen writers of the test, whose agendas were unknown, and the questions, therefore, difficult to predict. Not only can the symbolic psychic register be discerned, in which the "Big Other" of the test itself—its questions, its instructions and its statistical results—both allowed and constrained teacher actions, but there is also a sense of the productiveness of discursive practices. The teachers were produced as coaches through the apparatus of the test, kept in the dark about the test content, excluded from the process of test design, and distinctly separated from governmental assessment authorities. This discerned separation created what the teachers described as "pressure" and "stress." This pressure arose not only from their stated fear of not having prepared the children sufficiently, and, therefore, having possibly "disadvantaged" them, but from an even greater fear—that of possible accusations that they were ineffective teachers of mathematics, should their students achieve poor results. Although they described their choice to spend significant class time practicing for the test as "helping" their students, it seemed that this practice was just as much about their views of themselves as teachers. Using Mendick's idea of identity as a process of enactment, rather than a state of being, we can see Jemma at the cutting edge of such identity work, because this was her first experience of the test and she was unsure of what might constitute good teaching in such a situation. She described having given her students plenty of practice with number concepts, based on a belief that mathematics is predominantly about numbers. In an act of self-construction as a teacher of mathematics, she *recognized* this as the most important aspect of the subject. As it turned out, a higher proportion of the questions than in previous years were based on concepts from the geometry strand of the mathematics syllabus. Karen talked of this as a deliberate strategy to advantage boys. In her statement, Karen positioned herself, as a knowledgeable and experienced mathematics teacher, as understanding boys as spatial thinkers. For Jemma, some lack of preparation was justifiable, because the test was able to provide her with a more reliable truth about the mathematics that the children *could and couldn't do* than it would have if she had given them more practice in what she termed *ideal testing situations*, such as sitting on their own with desks apart, as she articulated in the following statement:

> I didn't really have any major stress problems in my class, but the thing was that they were sitting on their own, because on Fridays [test practice days], I don't always move their desks separately, so the fact that they were sitting on their own, that the environment was completely quiet, it was an indication of what they can do and what they can't do. But, like Karen said, that was [just] for me to know where my kids are and where I can improve in an ideal testing situation, because very seldom do I ever set up my classroom, when I give them a test, do I set up that complete sitting on your own, absolute silence, no talking, stop, something to eat, and start again. So, in that respect, it was kind of interesting to see how they responded to that testing.

The teachers' conversations about preparation showed that in their identity work as good mathematics teachers, the test was viewed as much a test of their own mathematical competencies as it was of the children's. Both the

Year 5 and Year 7 teachers reported that they were sometimes struggling to determine the correct answers for test questions, as these excerpts show:

> **Allie:** Even in the angles [one] as adults we all had huge discussions about that didn't we? You know the one with the angles.
>
> **Dan:** It was worded really strangely, and the kids got really confused. . . . They had to do the "insides" [interior angles].
>
> **Kate:** Some of it was ambiguous.
>
> **Jemma:** I found a lot of the questions were trick questions, rather than just doing straight operations. I was thinking that these twelve-year-olds were asked to do some horrendous problem solving.
>
> **Ralph:** I enjoy maths but I didn't enjoy all those spatial questions, probably because I hadn't covered as much of that type of stuff. . . . It's testing comprehension rather than mathematical skill.
>
> **Lee:** If we teachers don't know, how are the children expected to know?

The need for clarification led to discussions among the teachers to establish the correct answers. Within this teaching cohort, good mathematics teachers were discursively constructed as mathematical experts. It was decided that questions that were too difficult for the teacher must also be too difficult for children. To maintain (self)perceptions of mathematical competence, teachers had looked to external causes, such as "strange wording," "trick questions," "horrendous problem solving," or "ambiguity" to explain their personal difficulties in answering the questions since they were not "straight" operations. While it may have crossed their minds, none of the teachers seemed prepared to risk the ever-precarious security of their identities as mathematical experts (in a Lacanian view) by openly suggesting that their pedagogical content knowledge may have been insufficient. The power of the test to create (unexplained) "right" and "wrong" mathematical responses, and, hence, good and bad teachers of mathematics as knowledgeable or otherwise, appeared to produce a kind of identity angst for the teachers, which they attempted to resolve collectively. Teacher identity work can thus be seen within regimes of standardized testing to involve the adoption of localized self-protective strategies.

Another issue that emerged from the teachers' conversations was the perception that the test questions were not testing the children's mathematics, but rather their understanding of the wording of the questions. In the teachers' view, the test results gave a false reading of the children's mathematical capabilities. The following exchange was as much about the nature of mathematics and the ways the teachers produced themselves as teachers of this subject, as it was about the test itself:

> **Ralph:** [O]ne in every 35 mobile phones is tested, and it said, "There were 12,200 mobile phones made in a day, how many did they test?" It worked out to be...
>
> **Jemma:** It wasn't an even number.
>
> **Ralph:** It was 398.7, say, and so kids rounded it up. When you get a decimal number, you usually round it up. The process should have been one in every 35, you divide that number by 35 and get your answer. And a lot of them rounded it up, because that's generally what you do. It meant you had to go down... [because the question was asking for the number of phones].
>
> **Jemma:** A lot of them go, because it was over point five, it was point six, so you have to round it up, and in actual fact...
>
> **Ralph:** That's just an example. I think that that question could have been solved by... It's testing comprehension, rather than mathematical skill.

This conversation revealed the teachers' view of a *mathematical skill* as something that could be separated from its application in a meaningful context. They expressed annoyance that the test question had mislead the children, since they had been teaching them how to apply a rule for rounding that was "usually" or "generally" rounding up for parts of 0.5 or greater. In taking this line, they were explaining their role as one of presenting children with the mathematical rules and procedures that they were likely to need to answer predictable questions, rather than making sense of realistic and perhaps non-routine situations. We can see that the unexpected mathematical content of the test constructed the teachers, and mathematics itself, in a discursive process of expectation enacted in everyday practice. As far as the teachers were concerned, their job was to teach the "usual," not something that required interpretation in context ("comprehension"). The teachers were attempting to reconcile their personal (psychic) beliefs about what constitutes a "real" teacher of mathematics with the disturbing/ redefining vision the test presented. Such a conversation is an illustration of identity work in progress, the teachers engaging in a process of self (re) construction within the colliding discursive expectations and agendas created by the test. While the test created potential for the teachers to revise their view of the work of the good mathematics teacher, the top–down and high-stakes nature of the test regime did not appear to afford safe spaces for such professional readjustment.

According to the teachers, the test had a significant bearing on the way they delivered classroom mathematics programs. For some, it was seen as intruding on what they viewed as their proper role in deciding what mathematics the children should learn, and how:

> **Col:** To me, it rearranges my teaching format.... I used to do chunks of things [units of work], so, you know, in the fourth term [it] might have been the measurement or it might have been heavy in that, and, therefore, you've changed your whole way of teaching because you've got to do bits of everything.

Col's identity work as a teacher of mathematics could be seen at play in his articulation of his belief that a pedagogical approach where mathematics is taught in "chunks" achieves thoroughness and depth. The belief that learning mathematics should be taught thoroughly, rather than in bits, was also reflected in Allie's comment:

> In a way, this term has been totally modified, because you've been trying to teach the kids the things that they need to know for the test that's in August. You're disadvantaging them by not doing it. It has really altered, you know, almost like a whole term of nothingness....

Col and Allie were speaking of themselves in this exchange as professionals conferred the responsibility—and the freedom—to design and implement mathematics programs arranged in substantive units of study. The test created dissonances by reconfiguring this teacherly function. Teaching children "things that they need to know for the test"—teaching *to* the test, in other words—while rationalized by these teachers as ensuring the children were not "disadvantaged," was perceived as reducing their teaching to "nothingness" or "bits of everything." The test disrupted not only mathematics teaching programs, but the process of teacher identification, at the same time.

The test ran counter to what the teachers said they viewed as best practice in teaching and assessment of mathematics:

> **Yolanda:** I find it frustrating that we test them that way, when it's not what we teach. We teach them to talk and discuss, to find information, that's how they work now.... There's no way you can have it all in your head.
>
> **Karen:** When we are teaching, you get a mark for the formula and a mark for the process and a mark for the answer. The answer is the least important part, whereas understanding the process is more important.

Pedagogies of mathematics, as espoused in contemporary curricula, were reflected in these teachers' descriptions of their own teaching, such as, "we teach them to talk and discuss" and "understanding the process is more important." Speaking as those who knew how mathematics should be taught and learned, the teachers were able to criticize the test for its undermining of exemplary practice. In this way, the test compelled the teachers to define and clarify their pedagogical positions, while, at the same time, exposing the compromises they felt they must make in order satisfy the demands of the test. Jemma, on the other hand, expressed the belief that preparation for the test must be built into the Year 7 mathematics program, as part of best practice:

> I did mark the maths [before sending the test away to be officially marked], and I found some areas where the kids were falling down, and I felt, well, that's a weakness where I need to do some work, so I found it useful for me, personally, and I would like.... We all do as Year 7 teachers, maths tests, supposedly do them every week, that's part of our maths planning that we do a maths test at the end of each week, or, at the very least, every fortnight.

In this statement, Jemma referred to the test as something that could provide her with reliable information about children's "weaknesses"—a cause of their "falling down." She saw her job as repairing such deficiencies and monitoring children's achievement through a regime of regular testing as something she believed all Year 7 teachers were expected to undertake.

In other statements, the teachers spoke of themselves as intermediaries between the children, the test authorities, and parents. They were particularly concerned about the reaction of the parents; their reputations as competent teachers were seen to be under threat if the marks did not match parents' expectations. The teachers had developed a self-protective strategy for such a contingency—the role of the mediator. Both children and parents were told by the teachers that the test "means nothing":

> **Karen:** Parents will come in and say, "Can you explain this to me?" The biggest, most important thing to say to them is, [their child] may have got very close to the correct answer, they may have actually known what they were doing, but if they colour in the wrong bubble, they are wrong.... He understands, he knew to divide, but he made a "boof-head" mistake.

The teachers talked of their own assessment of the children's mathematics as more accurate, indeed, more "truthful" than that of the test. This placed them in a difficult position when the test results were found to be at odds with their own judgments of children's mathematical capabilities:

Dan: I have a much better idea [of the children's mathematical abilities] myself, of what I see every day. It's not a true reflection. I suppose it's a good way . . . the next reporting system to go back to parents to show where their kids are up to. I don't think it's totally accurate, so is that a good thing?

Karen: Sometimes it is a shock. If I think it's no reflection on the child's ability, I'll just tell them that. I had a kid with full-on flu sitting the test. So, when he got his test back, he was crying, he was so upset—and he was so bright. I said, "Don't worry about this test, mate."

As professionals, the teachers seemed to be caught between seeing the merits of the test as a "good way" to report children's achievement to parents, because of its supposed objectivity, and their view of the test as an invalid form of assessment. They trusted their own judgments, because they were based on professional observations of children's mathematical skills that the test could not "see." However, the underlying notion that a truth about the children's mathematical skills was there to be discerned by a trained and experienced professional was reinforced, rather than disturbed by, the test, as shown by Karen's use of the term "bright" to describe one student.

Teachers' visions of what it is that constitutes the good mathematics teacher were shaped and tempered in light of the test, as shown in the teachers' reflections:

Jemma: That's a concern of mine, that [the results] will come back and they will be down in the dark zone [*referring to the shaded part on the scale of results that indicates results below the state average or national benchmark*]. I think I do a good job, but I don't actually know. Maybe I'm not doing a good job. We know that there are some teachers around that really aren't good teachers, and this test would probably pick that up. If my kids don't do well I would have thought, "What am I doing wrong?"

Karen: It's not meant to judge the teacher. . . . Those kids bring so much to the test anyway, so if it was a really bad teacher, those kids could still do well.

Ralph: I guess the pressure comes in the results. I was confident that I prepared them well for it, but I haven't had to sit down next to somebody [*another teacher*] and compare my class and see whether my kids were above, average, or below, compared to other classes.

Jemma's distinction between "thinking" and "knowing" that she was a good teacher was telling. She expressed the belief that it was insufficient for her to simply *think* she was a good teacher. Only an external measure, such as the test, could tell her for sure. A Lacanian reading sees the test as serving as a reliable gauge of performance for Jemma, authorizing her imagined identity as real. As these comments show, it was worrying for teachers when the children's marks on the test differed from those they received in classroom-based assessment tasks, due to the tension produced between their *thinking* and *knowing* they were good teachers. Discrepancies in assessment results were also difficult for children and parents to reconcile. The teachers were then faced with a choice between explaining away the test results as invalid, or admitting they were not good teachers. It was a relief when the test results were consistent with their own judgments, since their desire to be seen as good teachers was fulfilled, as Allie's remarks illustrate: "[The test results] came out pretty much exactly the same as the record cards. It's nice when that happens. Normally, they look much worse. Maybe I'm getting better at it."

Allie's comment, "Maybe I'm getting better at it," suggested that she believed that a good teacher is one who is able to make judgments about children's mathematical abilities that closely align with the official, objective, standardized, and, therefore, more "truthful" judgment of the test. Ralph contemplated "good teaching" in a changing political climate, where such judgments might have wider implications: "Another point is, when it comes to the federal government's interest in performance-based pay . . . , a national standardized test, something like that, will be a guide. . . ." Karen expressed horror at the thought of test results being used as a determinant of performance-based pay, considering what this might do to mathematics teaching and, by implication, her core work as a teacher of mathematics: "Can you imagine how we would be teaching? We would be teaching straight multiple choice. Oh my God, it would be so boring." For Karen, the test had both personal and professional dimensions, as her identity work as teacher and mother became intertwined:

> I've got five children, and my youngest son, he sat the test [this year]. He's really quite bright, but I've gone home every year, I've done this for long a long time now, and I've gone, "Oh my God, stupid Year 7 test people, so rubbish, I don't really care." And yet, I've got some really bright kids, so they all achieved really well on the test. He went into the test, he didn't even do the circle [*coloring in the selected multiple choice answer bubbles*], he just went, "Yah . . ." [*mimes sitting doing nothing with arms folded*]. I said, "You dope," and he went, "But you said it was a stupid test."

Describing the test as rubbish, but at the same time taking it seriously in her capacities as teacher and mother, Karen could be seen in conflict in her accounts. Karen's attempt to resist the test's *telling her who she was*, in her invocation of personal agency, points to the challenges faced by individual teachers within institutionalized practice. Karen was caught on the uncomfortable horns of a dilemma: She would like her son to have been identified by the test as "bright," as had a number of her students, but she was convinced, as her long-held stance towards the test suggested, that the test did not tell a believable truth. Her identity work as "good mother" and "good teacher" were simultaneously engaged and mediated by the test.

DISCUSSION AND CONCLUSION

Standardized mathematics tests had become for these teachers an identifier, a signifier, and a *point de caption* in their ongoing identity work. From the teachers' remarks, there is powerful evidence that the test both challenged and reinforced their existing perceptions of themselves as teachers of mathematics. The teachers spoke of the format of the test as undermining the (real) methods by which they usually taught mathematics. As we could see from *The Guardian* excerpts, standardized testing is criticized for the ways it constrains and channels teachers' actions, particularly where classroom practice becomes dominated by "teaching to the test." There is clear evidence that, despite their recognition of process and understanding as important in their teaching of mathematics, this cohort of teachers treated the test—judged on children's answers only—not simply as a measure of the children's capabilities in mathematics, but also of their own teaching of the subject. Even though school policy dictated that a child's progress should be monitored and reported using a suite of assessment results gathered over time and using "authentic" assessment methods, the capacity of the test to define, shape, and reflect teacher identity subverted the school's attempts to offset what teachers described as the untrustworthiness of the test results. It also undermined mathematics teacher educators' attempts over the past two decades to change pedagogies of mathematics to include open investigation within communities of learners.

This analysis shows that the identity work involved in teachers' engagement with the test is complex. "Teaching to the test" can be seen as a response that is tightly tied to teachers' notions of themselves as professionally capable. Caught between multiple views of themselves as teachers of mathematics within the defining apparatus of the test, the teachers struggled to reconcile their beliefs about best practice, about their desire to be seen as "good" teachers (as indicated by positive test results). In the end, despite its disruption to teachers' mathematics programs and the difficulties it created for teachers in mediating between parents and children, the test was vested with authority as much from teachers' choice to go along with it for the part it played in their "identification" as teachers of mathematics, as from the intrinsic power it wielded as an externally imposed form of assessment. In the view of identity as work unfolding in narrative, such a choice produces the teachers, rather than being produced by them; in other words, the teachers were "choosing" to behave only in the teacherly ways that the test allowed and/or demanded, that is, as trainers and mediators between child, test, and parents. From a psychological perspective, the choice was tied to the teachers' desire (and need) for secure identities. While they may have railed against aspects of the test for the troubles it caused, in so doing, the teachers were producing themselves as professionals, as mathematically competent experts, and as good mathematics teachers. The test symbolized a commanding authority, made them "real" as mathematics teachers, and, at the same time, opened spaces for their visualizing of themselves as ideal teachers in ideal mathematics classrooms.

This research illustrates how teacher identity is as much about the social and psychical aspects of relationship, interaction, and techniques of power as it is about local needs, specific events, or specialized knowledge. This suggests that standardized mathematics testing is only marginally concerned with the teaching and learning of mathematics. It primarily serves, intentionally or not, to establish and maintain an accepted "order of mathematical identity," of which teachers' identities, in a continuous process of formation, are a (self)recognized and recognizable part. The more teachers of mathematics are bound by expectations that they teach to and administer standardized tests to account for their effectiveness as teachers, the more captive they will be through processes of teacher identity work, to a dereliction of what Ofsted identifies as good mathematics teaching. To return once more to *The Guardian*'s report:

> The government's testing regime...narrows children's understanding, and...leaves some pupils unable to explain mathematical theory because they are too used to answering narrow questions in tests....Pupils rarely investigate open-ended problems which might offer them opportunities to choose which approach to adopt or to reason and generalize. Most lessons do not emphasise mathematical talk enough; as a result, pupils struggle to express and develop their thinking.

Good teachers of mathematics are suggested here as those who foster pupils' understanding of mathematics through open-ended investigation engendering multiple approaches, discussion, reasoning, and generalization. Standardized testing policy is directly blamed for teachers' abandonment of this kind of teaching.

This research provides evidence to support such claims, suggesting that education policy makers who genuinely wish to improve the quality of pedagogies of mathematics and outcomes for children must closely attend to teaching as a process of identity work in which standardized testing plays a major part. They must address the ways in which standardized national testing has the potential to narrow not only children's mathematical understanding, but also teachers' identities as "good" teachers of mathematics. Teachers need, more than ever, the support of discourses that enable them to both imagine and recognize themselves, through the establishment of widely accepted and officially sanctioned symbolic representations of the good mathematics teacher, as a teacher who is *inspired*, who *emphasizes mathematical talk*, who acts as a participant in *open-ended investigation* in which *choice*, *reasoning*, and *generalization* are an accepted and everyday part of what they do, and correspondingly, who they see themselves to be.

The key theoretical ideas outlined in this chapter are useful for the ways in which they help us to interrogate and make sense of mathematics pedagogy in practice. A focus on teacher identity and the discursive production of "the good mathematics teacher" helps us to understand that teacher identification practices reflecting and embracing pedagogies of understanding, built through open investigation in mathematics, will only become possible with the development of appropriate discursive practices to support a reframed production of the good mathematics teacher. This could well mean rejecting timed, standardized, multiple choice answer tests, in favor of assessment that promotes and allows for substantive mathematical exploration and learning with understanding. As long as the public, education policy makers, and teachers themselves, as the implementers of policy at the classroom level, accept standardized tests as an unchangeable fact of life, it is inevitable that we will see more of the "bad" mathematics teaching that Ofsted deplores. Shaped by constraints and dilemmas produced in test preparation and administration, teachers will strive to reconcile multiple psychic and social demands in which they produce themselves through processes of identity work as "good mathematics teachers."

REFERENCES

Boaler, J., Wiliam, D., & Zevenbergen, R. (2000). The construction of identity in secondary mathematics education. In J. Matos & M. Santos (Eds.), *Proceedings of the Second International Mathematics Education and Society Conference.* Universidade de Lisboa.

Côte, J. E., & Levine, C. G. (2002). *Identity formation, agency and culture: A social psychological synthesis.* Mahwah, NJ: Lawrence Erlbaum.

Curtis, P. (2008). OFSTED criticizes maths lessons. *The Guardian*, 19 September, p. 23.

Foucault, M. (2002). *The archaeology of knowledge.* London: Routledge. (Original English translation published 1972).

The Guardian. (2008). Editorial, *The Guardian*, UK: Not satisfactory. 19 July.

Hall, S. (1991). Old and new identities, old and new ethnicities. In A. D. King (Ed.), *Culture, globalization and the world-system.* London: Macmillan.

Hogden, H., & Johnson, D. (2004). Teacher reflection, identity and belief change in the context of primary CAME. In A. Millett, M. Brown, & M. Askew (Eds.), *Primary mathematics and the developing professional.* Dordrecht: Kluwer Academic Publishers.

Holland, D., Lachicotte, W., Skinner, D., & Cain, C. (1998). *Identity and agency in cultural worlds.* Cambridge, MA: Harvard University Press.

Lacan, J. (1977). *The four fundamental concepts of psychoanalysis.* London: The Hogarth Press.

Lacan, J. (2002). *Ecrits: A selection.* London: W.W Norton & Co. Inc.

Mendick, H. (2006). *Masculinities in mathematics.* London: Open University Press.

Miller Marsh, M. (2002). Examining the discourses that shape our teacher identities. *Curriculum Inquiry, 32*(4), 453–469.

Schifter, D. (Ed.). (1996). *What's happening in math class? Volume 2: Reconstructing professional identities.* New York: Teachers' College Press.

Van Zoest, L., & Bohl, J. (2005). Mathematics teacher identity: A framework for understanding secondary school mathematics teachers' learning through practice. *Teacher Development, 9*(3) 315–346.

Wenger, E. (1998). *Communities of practice: Learning, meaning, and identity.* Cambridge: Cambridge University Press.

FURTHER READING

Beauchamp, C., & Thomas, L. (2009). Understanding teacher identity: An overview of issues in the literature and implications for teacher education. *Journal of Education, 39*(2), 175–189.

Cooper, K., & Olson, M. (1996). The multiple "Is" of teacher identity. In M. Kompf, D. Dworet, & R. Boak (Eds.), *Changing research and practice* (pp. 78–89). London: Falmer Press.

de Freitas, E. (2008). Enacting identity through narrative: Interrupting the procedural discourse in mathematics classrooms. In T. Brown (Ed.), *The psychology of mathematics education: A psychoanalytic displacement* (pp. 139–155). Rotterdam: Sense Publications.

Foucault, M. (1970). *The order of things: An archaeology of the human sciences.* New York: Random House.

Foucault, M. (1977). *Discipline and Punish: The birth of the prison.* London: Allan Lane.

SECTION 2

DISCURSIVE APPROACHES TO PEDAGOGY

CHAPTER 5

FRAGILE LEARNING IN THE MATHEMATICS CLASSROOM

How Mathematics Lessons Are not Just for Learning Mathematics

Diana Stentoft and Paola Valero

ABSTRACT

The "strong social turn" in mathematics education research (Lerman, 2006) has brought cultural, political, and social issues into close contact with what it means to teach and learn mathematics. In our work, we have been challenged by ideas, not in relation to the production of narratives of truth about the social world, but rather with the formulation of alternative windows to look into the practices of education, in particular, the highly socially valued practices of mathematics education. In this chapter, we discuss one central contention: When a diverse group of students in a classroom engages in mathematical learning in the classroom, they bring with them multiple resources of identification, based both on their previous experiences as well as on their future aspirations. Shifts in discourse continuously bring to the fore a variety of students' identities as students engage in learning not only as "mathematics learners" or "future mathematics teachers," but also as "football fans," "music

Unpacking Pedagogy: New Perspectives for Mathematics Classrooms, pages 87–107
Copyright © 2010 by Information Age Publishing
87

enthusiasts," "family members," and so forth. We capture the shifts in discourse and in identification by characterizing discourse, identity, and learning as inherently fragile. The fragility of the classroom discourse, of learners' identities, and of the learning processes allows us to propose that mathematics classrooms are sites not simply for learning mathematics. We believe our contention demands serious reflection and actions from mathematics teachers and in mathematics education research.

INTRODUCING IDEAS

For us, Diana and Paola, venturing into poststructuralist research means an acknowledgment of the uncertain, the ambiguous, and the fragile as a premise of mathematics education, as well as research.[1] A first important tool we adopt is the overall idea that a poststructuralist and postmodern[2] perspective bears an attitude of critique to what has been taken for granted (Valero, 2004). More precisely, this critique allows us to transgress established ways of seeing and understanding practices (Biesta, 2006) in a search for the impossibilities, disturbances, and hidden potentials within the established order. In our poststructuralist analytical move, we shift the focus from what seems to be central to what seems to be ignored and excluded. This move allows us to address and phrase the counter-practices existing alongside with, for example, practices established in response to formal educational requirements. Biesta (personal communication) proposes that a poststructuralist move is an immanent strategy that brings into focus what other views (particularly modern views) would construe as "noise." For example, if research focuses on what is inside a square, a poststructuralist research move rotates the square, so that some areas that were not previously in focus become part of the research gaze (the shadowed triangles). Of course, some other areas (the small grey triangles) will now be invisible to the research gaze. In this sense, poststructuralist research also creates its own "noises" (Figure 5.1).

We found a second tool to underpin our analysis, and that was the view that all constructions are language games. Lyotard (e.g., 1984) proposes language games as a significant constituent of the social bond. He argues:

> The self does not amount to much, but no self is an island; each exists in a fabric of relations that is now more complex and mobile than ever before. Young or old, man or woman, rich or poor, a person is always located at "nodal points" of specific communication circuits, however thin they may be. (Lyotard, 1984, p. 15)

We can use this notion of language games to demonstrate how impossible it is to take for granted the idea that the framing of an educational

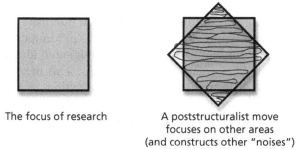

The focus of research

A poststructuralist move
focuses on other areas
(and constructs other "noises")

Figure 5.1 Illustration of a poststructuralist research move.

institution is sufficient to construct students as learners. Acknowledging the unpredictability of language games and recognizing all its possibilities inevitably challenges perceptions of the social order and shifts attention from overarching structures of institutions and educational policy to the locality of discourse, interaction, and identity construction.

Working from a hypothesis of the fragile and the non-committal state of individuals in an equally negotiable society opens us up to new perspectives on how to perceive teaching and learning mathematics. Using these perspectives, we contend that learning cannot be considered a stable, predictable, and repeatable process, but rather must be seen as a fragile, intermittent, and discontinuous process, vulnerable to the interrelated and continuous constructions and alterations of discourse and identity. Offering this view on learning, we identify some underlying assumptions. First, an individual's identity construction and participation in discursive practices are deeply interrelated, are constantly changing, and, thus, they constitute the fragile and unpredictable environment in which learning emerges. Consequently, it is difficult to imagine discursive practices in the mathematics classroom that are purely focused on talks and interactions relating to mathematics and that are sustained over an extended period of time. Interactions and activities that can be clearly identified as mathematical have only a limited duration in the span of a lesson. The instability of discursive practices and identities is not a signal of a "bad" or "poor" mathematical learning environment—an interpretation that has been frequent in studies examining mathematics classroom communication and discourse. It is simply what can be expected to happen when human beings—teachers and students—engage and participate in mathematics classroom interaction. Second, we foreground interpretations of a "noisy" classroom—a classroom where learning mathematics is but one possibility that may be explored or rejected in conversation with the discursive practices and identities that are continuously brought into play in classroom interaction. Third, the learning potential of students is closely connected to the processes of partici-pants constructing and shifting their identity in discursive practices. What an individual can learn is closely related to his/her immediate and actual identity, as well as his/her imaginations or designated identity about his/her future possibilities (Sfard & Prusak, 2005). This means that in class-room discourse, for example, a person who identifies as a football player will find conversations about football relevant, but may see no point in en-gaging with discursive practices that could lead to learning mathematics. However, identities and discursive practices are constantly changing, and the transition from identification as a football player to identification as a mathematics learner may be relatively smooth.

Bauman (1996) has described the postmodern man as a vagabond or tourist seeking to untie himself from life's restrictions. Where the man of the modern era sought to build an identity, the man of the postmodern has as his mission not to commit himself to any particular identity over prolonged periods of time. We agree that identity should, therefore, not be seen as a lifelong project of any individual, and in this it is impossible to see identity as serving a role of informing both the person himself and the world around him about his core, his inner-most, and who he is. Instead, identities should be referred to in plural. They are not fixed and structuring sources of infor-mation, but are, rather, flexible and temporal attachments of the individual to the discursive practices of which he is part (Hall, 1996). Hall proceeds to suggest that "identities are never unified and, in late modern times, increas-ingly fragmented and fractured; never singular but multiply constructed across different, often intersecting and antagonistic, discourses, practices and positions" (1996, p. 4). The fragmented state of identity and discourse, and their continuous and mutual feeding of each other largely render fixed categorizations and extra discursive layers prescribed by research or society without much meaning. We might ask, "How might categories and identities be accurately predicted and prescribed in research and how might their stability be ensured through time and space?" Or, put differently, "Is the teacher always a teacher and is the student always a student?" In posing these questions, it becomes apparent that a poststructural perspective may have considerable implications for how we can frame a notion of the learner and how can we characterize the processes of learning.

In most research, the concept of learning is closely connected to as-sumptions of an unambiguous process separable from other processes. It is assumed that processes of learning can be analyzed and scrutinized in-dependent of the context in which the learning is intended to take place. It is asserted that learning is an individual cognitive affair or a product that the individual receives in the form of knowledge transfer set within formal structures of education (Fuller, 2007). A similar picture can be painted in the field of mathematics education, where, although "noise," in the form of socio-cultural and socio-political influences on processes of learning and practices of teaching, is recognized to an increasing extent (Lerman, 2006), questions are rarely posed in relation to the constitution of mathematics classroom interaction, or to that which does not contain any obvious relation to teaching and learning of mathematics. We find exceptions in the work of Vithal and Valero (2003), who, in their explorations of situations of conflict in mathematics education, call for research to take into account "the whole" school or "the whole" classroom (p. 559), even when researching specific aspects of learning mathematics. By "the whole," they refer to anything social, political, or conflictual, which, they argue, can bear significantly on the possibilities for learning mathematics. This "whole" naturally also includes the "noise," or non-mathematical interaction, of the mathematics classroom. Lange (2007) also specifically addresses non-mathematical aspects of classroom interaction in his research on children's perspectives on mathematics education, and, specifically, on how issues of friendship are interwoven into children's mathematics performance and their constructions of meaning of mathematics. Alrø and Skovsmose (2002) have proposed zooming in and zooming out of situations of learning. In their analyses, they have explored how students show resistance to learning mathematics and, instead, produce disturbances and "noise" in the mathematics classroom. We will argue that conscious resistance to learning mathematics is only one of many potential sources of "noise" in the mathematics classroom.

In shedding light on learning from a poststructuralist position, we have come to see learning not as an unambiguous process with clearly defined borders in relation to contexts, participants, and other disciplines, but, instead, as an action performed in and through discursive practices and strongly connected to the immediate identity of the learner. Hall (1996) argues:

> Precisely because identities are constructed within, not outside, discourse, we need to understand them as produced in specific historical and institutional sites within specific discursive formations and practices, by specific enunciative strategies. Moreover, they emerge within the play of specific modalities of power, and thus are more the product of the marking of difference and exclusion, than they are the sign of an identical, naturally constituted unity.... (p. 4)

When considering the mathematics classroom, we can thus see the identities constructed in and through discursive practices not as random, but rather as products of past experiences and imaginaries about the future, as well as the present, in the classroom setting. Constructions of identities can, therefore, be seen as complex acts, which further points to the problematic in taking identities relating to, for example, gender, race, or ethnicity for granted when undertaking research into the learning of mathematics. A poststructuralist perspective on the notion of identity strongly opposes a view on identity as a static and constant entity. Baumann (2004) links identity and belonging to peoples' actions, and proposes both as fragile and dynamic:

> One becomes aware that "belonging" and "identity" are not cut in rock, that they are not secured by a lifelong guarantee, that they are eminently negotiable and revocable; and that one's own decisions, the steps one takes, the way one acts—and the determination to stick by all that—are crucial factors of both. (p. 11)

A poststructuralist perspective on identities and learning comes with some distinguishing characteristics. First, we propose learning to be invariably located in a world, a society, and an individual, who is faced with constant ambiguities, in terms of process, as well as content. There is no universality to set the frame or the standard for the learning process. Consequently, we see knowledge as local and multifaceted (Kilgore, 2004). Second, we see learning as being contingent on immediate discursive practices and identities that, together with the intersecting life trajectories, past experiences, and future aspirations of participants, constitute a landscape of learning. In the words of Deborah Kilgore (2004), "The postmodern learner is always becoming, always in process, always situated in a context that is also always becoming" (p. 47). Thus, the constitutive parts contributing to the landscape of learning are fragile. However, this is not to suggest that learning is random, rather, that the potentials for learning are local emergences that cannot be accurately predicted through curricula, text books, or standardized teaching strategies. Sfard and Prusack (2005) propose learning as the gap between the *actual* (present) and *designated* (imagined for the future) identities of any individual. In this scope, we view learning as the process moving the individual toward his future aspirations and toward becoming, though bearing in mind that both actual and designated identities are of a fragile and ever-changing nature (Stentoft & Valero, in press).

Tying learning so explicitly to fluid identities of the individual, we propose a notion of learning not as a process similar for all, but rather as a process unique to the individual and inextricably linked to his or her being in relation to discursive practices and identities and to his or her life trajectories. Seeing learning as an individual affair, incessantly dependent on the social, calls for a shift in classroom focus, from defining and designing optimal methods of instruction inclusive and appealing to all learners, toward an increased attention on interactional processes, as they appear in the classroom, in which an environment and learning situations that are meaningful to any one individual are created (Klein, 2002).

Figure 5.2 illustrates elements and processes in a landscape of learning, and points to their continuously changing intersections and interactions. As was argued above, we do not see the notion of learning as independent, or as a strictly individual or strictly social process. Instead, we see learning as

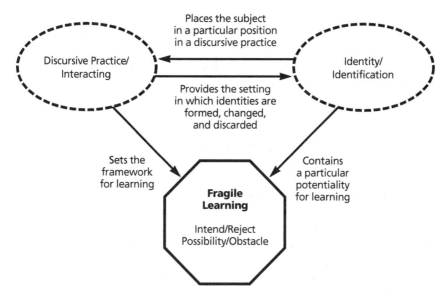

Figure 5.2 Representation of the landscape of fragile learning.

being in close interplay with the complexities embedded in ever-changing discursive practices and identities (Sfard & Prusak, 2005). Students' fragile identities can be seen as playing a significant role in the mathematics classroom, as they may contain potentialities for learning mathematics, thus making this particular kind of learning a possibility. On the contrary, they may not be connected to any potential for learning mathematics, in which case, they can be seen as learning obstacles.

Recognizing a notion of learning as fragile, and discursive practices and identities as constitutive components and determinants of the nature and content embedded into the learning process presents some implications for how we can conceive of mathematics classroom interaction. First, the continuous possibilities for changing discourses and identities means that learning mathematics and constructing mathematical knowledge cannot, despite the formal structures of a mathematics classroom, be taken for granted. The prevalence of an identity as a mathematics learner cannot be assumed, but lies inherent in some discursive practices, and not in others. Second, from the fragility of learning naturally follows that the construction of mathematical knowledge is also fragile. Instructional materials, teacher authority, and the institutional setting of a classroom are all tools provided to accentuate specific discursive practices and identities that favor particular constructions of mathematical knowledge. However, participants' actual intention to learn mathematics still rests on their immediate and continuously interchangeable identities.

In the following pages, we present the outcome of a case study into mathematics classroom interaction. The aim of the investigation was primarily to uncover the significance of "noise" in a mathematics classroom, as it emerged through ever shifting identities and discursive practices, and to explore how this "noise" contributes both to students' possibilities and obstacles in learning mathematics.

APPLYING IDEAS

The material we present below is part of a case study produced in the classroom of a Danish teacher training college over a period of eight months. We report on one month of the study. The students were enrolled in a teaching degree (B.Ed.) that involved an evening program, allowing them to simultaneously hold daytime jobs. The age and background of the students varied dramatically, with some coming straight out of secondary school, some re-educating after several years of employment in various fields, and some having completed a one-year preparatory course for non-Danish students wishing to become teachers in the Danish Folkeskole (serving students of approximately 6–16 years). All the students had chosen to specialize in mathematics, as one of four specialized subjects. The mathematics course was organized with four lessons in one evening every week. Prior to making the recordings of the classroom interaction, Diana organized meetings with the two teachers teaching mathematics and psychology/pedagogy respectively. These teachers had developed a strong interest in this particular class. They viewed students' diversity as simultaneously enriching and problematic. Some pre-service teachers experienced obvious problems, as their Danish language skills did not allow them to write or express themselves at a sufficiently high level. The training college teachers further pointed out that many students who do not have Danish as their mother tongue chose mathematics as one of their specialized subjects.[3]

Classroom Interaction and Discourse

The poststructuralist ideas expounded above set some methodological premises for the construction of the case study, the gathering and handling of empirical material, and the subsequent analysis. Researching from an epistemology of the local and the fragmented entails an acknowledgment of the analysis and research findings as representing local and permanently contestable knowledge of the researcher. As argued by Kincheloe (2003), it is impossible to separate the knower from the known; the researcher is, as any human being, a perceiving person and a reflexive subject in constant interaction with the world. In rejecting totality, the study did not have as its mission to uncover universal truths about the mathematics classroom, stable through time and space, or to ascribe normative features to what is named mathematics classroom interaction and students. Instead, the focus on the locality of knowledge and discursive determinants of the classroom interaction made naturally-occurring talk-in-interaction appear an appropriate source of information. Interviews or classroom observations, in this view, would simply be a researcher's interpretation of interactions of the past. Interview materials would be constructed especially for the event of the study and in discursive practices and participant identities far removed from the actual events of the mathematics classroom. The assumption of knowledge as fragmented, locally constructed, negotiated, accepted, and rejected, all in a continuous process of learning and contingent on fluid discourses and fragile identities, disqualified any use of pre-ascribed identities or advance categorizations of classroom participants as a basis for the study and the initial analysis. In the analysis, the "text" of the classroom was left open and un-coded to speak the voice of participants and let participants set the agenda for topics and themes of relevance. We should, however, note that we acknowledge that what appears in our analysis is inevitably the result of our prioritizing and of our own interpretations of classroom interaction. Hence, readers may notice the absence of classical identifications of race, gender, ethnicity, and so forth, in the analysis as strong contributors to the interaction or embedded into discursive practices. Although aspects of categories such as race or gender could inform classroom interaction through experiences and imaginations of participants, such identities were not dominant in the interaction.

The classroom interaction was recorded on audio-tapes on three nights, each containing four lessons of mathematics. For the sake of managing the technical side of recording the interaction, Diana was present during the recordings, but did not solicit any interaction and did not contribute to the teaching process. Subsequently, the recordings were converted into electronic files, which served as the foundation for the analysis. What emerged in this specific classroom is idiosyncratic to that classroom. Our objective is to address processes of interaction as a source of information when analyzing and to explore what the process of analyzing fragile identities and discursive practices might reveal, with regard to the learning of mathematics as a possibility or an obstacle.

Analyzing Interactions

The analysis of the classroom interaction presented numerous challenges related to theoretical assumptions and the criteria for selecting specific extracts for deeper analysis and presentation. The nature of interaction in a busy mathematics classroom made transcribing the material a complicated and untenable option, in which relevant utterances and sequences could easily be overlooked, misquoted, or would be difficult to follow. The alternative came in the form of a database in which entries were made pointing to scenarios, passages, discussions, and the like, where alternating discursive practices, identities, and learning were evidenced in the classroom interaction. The database did not serve as an extra layer of interpretation, but rather as a resource allowing the researchers to revisit the recordings of the classroom at particular points of the interaction. To this end, the database was designed with fields containing information about, for example, time of sequence, participants interacting in the sequence, description of the sequence, and a field for transcription of the particular sequence. The database proved a useful tool for organizing the content of the recordings in a way that allowed for the locality of events to remain in focus.

The analysis of the recordings was carried out twice, from two different perspectives. The first round of analysis was centered on self categorizations and social identities, as inspired by Turner and colleagues (1987) and Tajfel (1982). This analysis pointed to sequences of the interaction where participants' identifications were made explicit in the classroom. This analysis was, to some extent, organized according to who carried out the identification/categorization and whether these identifications had any connection to the learning of mathematics. Although indicating points of identifications in the classroom, this initial analysis did not focus on the interplay between interactions in and out of discourses of learning mathematics. In this, the analysis challenged the theoretical assumptions of categorization and their perceived potentials for contributing to a perspective on local classroom interaction.

The second analysis, which generated additions to the database, was primarily framed by a poststructural emphasis on process, rather than stable identifications. Revisiting the recordings from this perspective revealed three themes embedded in the processes of being and constructing in the mathematics classroom. The themes are: *fragile identities, fragile learning*, and *fragile mathematics*. The themes emerged in various discursive practices in the classroom interaction throughout the analysis. The themes of fragile identities and fragile learning are explored further on the following pages. The theme of fragile mathematics is complex and requires a somewhat different analysis, and therefore will not be explored in this chapter.

The Fragility of Identities and the Learning of Mathematics

The analysis revealed a constant alternation between discursive practices and identities in the mathematics classroom. The extracts presented below were selected as exemplary and particularly illustrative of these alternations and their consequences and potentialities for learning. The constant alternations in discursive practice and identities brought out in classroom interaction were made manifest through the continuous navigations of students and the teacher in and out of discourses relating to the learning of mathematics. Consequently, attention was given to extracts revealing alternations between practices of mathematics and practices of "noise" (or "non-mathematics") to focus in on the fragility of these practices and the continuous scope for alternations. Pseudonyms such as S1, S2, and T have been created for convenience and to preserve anonymity. However, it is important to realize that this is by no means to suggest these identities are in any way superior to others or remain constant through time and space.

Fragile Identities

As set out in the theoretical section above, it is difficult to imagine a mathematics classroom where all interaction is strictly related to the learning of mathematics, with no regard for variations in past experiences and imaginaries about the future and the potential for the emergence of different discursive practices. In a group work session where the participants are working together in pairs on a mathematical assignment, two participants who have been actively engaging in work on the assignment are heard discussing a building project. Meanwhile, other groups of students are heard continuing their involvement with mathematics:

Extract 1

> **S1:** Isn't there some . . . (inaudible)
> **S2:** (coughs)
> **S1:** You should know.
> **S2:** Yes. Well, then there is some, eh. . . . If you could, then you could do something different. Then you could put. . . . If this is a board.
> **S1:** Uhu.
> **S2:** A big board. Then you can put some sound-proofing material inside it. So even if it wants to vibrate, the . . .
> **S1:** Oh, yes . . . (inaudible)
> **S2:** Yes. Something called Dectason. . . . You could . . .
> **S1:** This thick?
> **S2:** Yes. You could put this inside.
> **S1:** Yeah.
> **S2:** If there is room.

The participants in this extract are discussing a situation far removed from the mathematics assignment, engaged in a conversation drawing extensively on discursive practices originating outside the mathematics classroom and outside the mathematics content. S1 has a problem he needs to solve in order to complete a project of sound proofing. He has a vision of the completed project and, in this, his images of the future are influencing the present discourse and his immediate identification as a builder. S2 draws on his past experiences with this kind of work and provides advice, identifying himself as being someone with building experiences. S1 and S2 are physically present in the mathematics classroom, but, in this particular instance, they can hardly be categorized as students, as they are not in a discursive practice allowing them to engage with the work on the mathematical assignment. Instead, S1 and S2 engage with different learning potentials as they involve themselves with solving a problem of sound proofing.

Football appeared as a recurring theme in the classroom interaction, in which the involved individuals drew on experiences outside a discourse of learning mathematics. Prior to the following sequence, T and S3 were discussing exam papers in mathematics and considering which questions and mathematical topics could be expected in the exam:

Extract 2

> **S3:** Yes, well you kind of need that frame of reference in place.
> **S4:** Yes, yes.
> **S3:** And, and, and (inaudible) . . . It can be very good to have, eh, an example. Right, kind of a clear example. I can remember there was something about some triangles. What. I don't remember because. . . . Well, yes, that. . . . There is so much information coming at the moment that, eh. . . . (Laughs) Do you remember the one about . . . ? (interrupted by S5)
> **S5:** Write to tell how it goes in the football match! (students laughing)
> **T:** I have looked into my crystal ball. Brønby wins, so you don't have to worry.
> **S5:** No, no. . . . Did you see them play this weekend?
> **T:** Um, when they lost 0–1?
> **S5:** Yes.
> **T:** That (mumbles) that was Saturday. No, I did not see that. I don't have that channel.
> **S4:** (mutters something about Brønby).

S5: What do you say? (laughs)

S4: Let's move on.

S5: As usual.

  T: Very well. That was one of the things I want you to be able to do, including S3's comments. The other thing is that I want you to understand the equations of the circle.

The sequence illustrates how a comment from one classroom participant shifts classroom discourse from talk about the mathematics exam toward football. T and S5 engage in talk about a particular match, and, momentarily, cannot be characterized as student and teacher of mathematics. Rather, they can be identified as individuals interested in football and with some experience associated with the match in question. S3's concerns regarding the mathematics exam are temporarily replaced by talk of football. In the discursive practice addressing football, mathematics is removed from center stage, as is mathematics learning potentials. The extract above and the study as a whole did not reveal how S3 received this shift in discourse and identities. The example does, however, illustrate how alternating identities and discursive practices of some participants carry the potential to interrupt or overrule the discursive practices of others. Eventually, the teacher re-assumes his identity as teacher, re-directing attention towards his intentions and objectives of the students learning equations of the circle.

The situations in which classroom interaction strays away from its institutionalized purpose of teaching and learning mathematics are numerous and occur throughout the study. In turn, the variations in discourses and identities and their embedded learning potentialities are substantial. Even within the institutionalized framework set in place to further the teaching and learning of mathematics, for example, in the form of the physical layout of the classroom, provision of teaching materials, or the regulations and curriculum specifying the content and outcome of the mathematics course, various identities and discursive practices prevail.

Learning mathematics as a specialized subject entails more than acquiring skills and knowledge in mathematics. The course also contains integrated elements of mathematics didactics intended to initiate participants' reflections and considerations of being or becoming teachers of mathematics. The study revealed numerous sequences of participants alternating between identities of learners of mathematics and what Sfard and Prusak (2005) label *designated identities* of future teachers. The teacher also engaged in discourses emphasizing participants' identities as future teachers of mathematics. This is visible in extract 3 when the teacher in his talk draws attention to didactical tools for teaching mathematics.

Extract 3

> **T:** It is of no use if S1 can solve equations much better than any one of you, if he cannot explain what he is doing. Fortunately, he can. But a substantial part of your task as teachers is not just to teach the kids to do mathematics assignments, but also to teach them how to explain. And it is a very important part of the linguistic development of the kids happening in mathematics lessons.

In the extract, the teacher orients himself toward participants as future teachers—teachers to be. He enters into an exercise of articulating a designated identity of future teachers, and simultaneously connects this imagination to discursive practices of engaging with mathematics. The teacher clarifies to the participants what is required of them in order to reach their aspirations for the future, and what is required is exactly in line with Sfard and Prusak (2005), when they argue that learning is the gap between actual and designated identities. Thus, in this instance, learning to explain mathematical principles is part of the journey participants must make in order to reach a destination of "mathematics teachers."

Fragile Learning

Identities are closely tied to learning, and in their fragility they render the learning process a fragile enterprise, vulnerable to variations in identities and discursive practices. Each identity contains an inherent potentiality for learning. However, the study demonstrates that neither the idea of mathematics as a constant focus nor the possibility of learning mathematics can be taken for granted. Instead, continuously changing discursive practices and identities in the classroom shape a multitude of different learning potentialities and, as will be demonstrated below, also serve as obstacles or barriers for learning mathematics.

Prior to the following sequence, the teacher had been demonstrating how to write the letter *b* on the blackboard in such a way that it would not be confused for the number 6. In this, the teacher addresses classroom participants as future mathematics teachers.

Extract 4

> **T:** ... But as soon as you have it on the blackboard, then you mix the number six and that way of writing *b*. And YOU can see the difference in your own writing, but the kids cannot.

Is it six or is it *b*? It is a small thing to be aware of. Now, I know, I can't remember how you write *b*, but if you do it in mathematics, then write the *b* like this, so it is open, or in some other way that cannot be misunderstood. Hi S6.

S6: Sorry I am late.

T: As long as you come.

S6: The trains were going wrong.

T: Did you see anyone else on the train?

S6: I don't know. But it was, um, there was a train jam in V station.

T: Aha. . . . Train jam.

S6: Yes, it was something about too many trains ending in V. So I did not get on the right train.

T: That's a wonderful word. Have you heard this word before—a train jam?

S7: Yeah.

T: Have you heard this before?

S7: Yeah, in my language.

S6: Have you heard it? . . . (inaudible)

T: Bosnian.

S7: Yes. S8 says it exists.

T: S8 says it exists?

S7: Yes.

Prior to this sequence, the teacher was talking about mathematical representation, and the students were actively listening and engaging in the discursive practice, appearing to pay attention and, in this, demonstrating a potential to learn about this mathematical representation. They were learning and moving towards a designated identity of future teachers of mathematics. This discursive practice is immediately transformed when S6 enters the classroom and explains her late arrival. A new learning potential emerges in the discussion about the word construction *train jam*. Does this phrase exist? Does anyone know? Yes, it does in the Bosnian language. The potential for learning remains, but now in a new discursive practice and new identities of being Bosnian, being a late-to-class student, and being a man with an interest in word constructions and their applicability (the teacher). The fragility of learning (mathematics) was equally evident in extract 2 above, where participants are seen to move in and out of discourses of mathematics, orienting the discourse and identities towards exams, football, and mathematics.

In their interaction and shaping of discourse and identities, participants incorporate their experiences and aspirations for the future as they offer contributions to the various discursive practices: being a mathematics student, becoming a teacher, being interested in football, exploring the meanings and existence of a word. And in their practices and interaction, participants continuously pursue potentialities for learning—about words, about mathematics, about football, and, not least, about themselves and each other.

In a conscious or unconscious attempt to guide or influence discursive practices and identities in the classroom toward practices containing a potential for learning mathematics, the teacher utilized various strategies to divert attention toward his identity of teacher and toward mathematical topics. Phrases of *get back to work!*, *I recommend . . . , Well!* are often made in a loud voice, as a means of interrupting whichever discursive practices are operating in the classroom. Extract 5 provides a clear example, when participants talk about football while one student is writing out a solution to a mathematical assignment on the blackboard. Throughout the sequence, the chalk against the blackboard is audible in the background.

Extract 5

> **T:** Yes, it is a long time since I stood by the sideline shouting "the referee is a carrot" (singing the words in a mocking tune).
> **S5:** (laughs) No one shouts that. (laughs)
> **T:** That, we used to in the early '60s.
> **S5:** It's called "scandal referee" (singing in a mocking tune).
> **T:** In Belgium, you shouted "Ref le spec!"
> **S5:** What does that mean?
> **T:** referee le spec—put on your glasses, then! Don't you have any glasses on? WOW, WOW, WOW!
> **S5:** Something happened already.
> **T:** Now we should remember!
> **S2:** Something has happened already.
> **T:** When you write $x - 10$.
> **S6:** Yeah.
> **T:** Times $x - 10$ that equals x^2 here $- 20x$, that is there.
> **S6:** Yeah.
> **T:** Then you have to say . . .
> **S6:** Yeah, but then I subtract and it is therefore. . . .

In the sequence, T is seen to shift between two different discursive practices and two identities. One locates him as a football enthusiast in his younger years, and the other as a teacher, abruptly introduced in response to the student at the blackboard when she makes a mistake in her writing of a solution to a mathematics exercise. In his outburst, *WOW, WOW, WOW!*, T is not only marking a shift in his own discursive practice. In the subsequent responses from S5, S2 and S6 it is revealed how they follow him into a discourse of mathematics when they remark that something has happened in reference to the writing on the blackboard. The shift of discourses and identities from football to mathematics marks an abrupt shift in learning potentialities—from a discourse about football slogans in the past to a practice of mathematical instruction. Again, it is seen how the focus on particular learning is temporary and can only be maintained through the maintenance of corresponding identities and discursive practices. Consequently, the only constant in the mathematics classroom is the potential to change and learn.

MATHEMATICS CLASSROOMS ARE NOT SIMPLY FOR LEARNING MATHEMATICS

Set within educational institutions are procedures that regulate the way in which individuals are expected to perform and engage with their education. These procedures are generally constructed to meet the demands of other institutional settings: producing teachers for primary and secondary schools, producing doctors for hospitals, producing administrators for business and public administration, and so forth. The settings of institutions are designed to favor particular discursive practices and identities, and reject others; they are what Gert Biesta (2006) refers to as rational communities legitimating particular institutional and educational discourses and particular voices, and not legitimating others.

The study presented in this chapter demonstrates how a class of pre-service teachers does not engage in learning mathematics solely at the rational level. Despite its formal setting in an educational institution, and despite its design as a rational community, the course included voices of illegitimate discourses and identities pertaining to football and other aspects of participants' being. Those voices were part of the overall agenda in the classroom and, in that sense, carried the potential for learning. We can point to significantly different strategies for dealing with fragile discourses and identities: either to accept and embrace their existence in the mathematics classroom, or to make even greater efforts to suppress their existence and maintain a focus on rational communities, engaging with mathematics. In the study, the teacher chose to engage with other discursive practices than those pertaining to teaching mathematics. However, on numerous occasions, he also sought to re-initiate exactly those discursive practices and identities containing potentials for learning mathematics.

Viewing learning as fragile and closely connected to discourse and identity entails, if not a legitimating of these discourses and identities, then at least an acknowledgment of their existence and presence in the classroom.

It shifts the focus away from the mathematical content—what a student (in his or her student identity) must learn—toward the processes setting the environment and framework for the learning of mathematics as a possibility among other learning possibilities. While it has never been proposed as a guarantee for participants to learn mathematics, the perspective presented in this chapter, as a point of departure, moves one step further in suggesting that mathematics teaching materials, a mathematics teacher, and the setting of a mathematics classroom do not guarantee that participants will engage with mathematics. In other words, there are obstacles pertaining to immediate discourse and identities to overcome before the mathematical content and learning mathematics will appear relevant. Neither students nor teachers can be expected to commit themselves exclusively to mathematics; they are unfaithful to an identity often assumed and taken for granted in both teaching and research.

While embracing a poststructural perspective, realizing the uncertainties of the mathematics classroom, it is still possible to establish at least two criteria for successful engagement with mathematics education, both of which pertain to processes within the social realm. First, discursive practices and identities must contain potentialities for the learning of mathematics. These potentialities are emerging as a result of the constant negotiation of discursive practices and identities, and cannot be taken for granted. As demonstrated in the extracts above, discourse and identities in the mathematics classroom are fragile and require nurturing, if one is to establish or maintain full classroom potential. It is even questionable whether a full classroom potential exists. Second, the potentialities for learning contained within particular identities are determined not only by the interrelatedness of discourse and identity, but also by the life trajectories of participants. Fragments of life trajectories are on constant display in participants' interaction with each other in the classroom, as was noted in the extracts about the building project and the football match, and the way in which they inform participants' responses to particular situations. Having experiences with the activity of building allowed S2 to provide advice on the building project and assume a temporary identity as a building expert in extract 1. Imagined life trajectories pointing to becoming a mathematics teacher were unfolded in extract 3. Setting these criteria forth does not in itself provide participants in the mathematics classroom with concrete tools for learning. However, they do offer explanations both for successful and less successful mathematics classroom interaction, where the explicit and institutionalized objective is for students to learn mathematics. They also offer a theoretical framework for participants to reflect on their own participation in various discursive practices and identities. This is in line with the analysis provided by Davis and Williams (2009), in which hybridities of mathematics talk and peer talk in the mathematics classroom are addressed. Hybridity is proposed as a strategy for creating a more vibrant and engaging mathematics learning environment (Davis & Williams, 2009). The proposal is strongly concerned with opening up learning potentialities by looking outside mathematics itself and into how students and teachers engage in multiple learning possibilities in their social interaction in the mathematics classroom. This suggests a need for re-directing attention from mathematics content and mathematically competent educators towards a broader socio-cultural and socio-political perspective. Such a perspective would entail a reconsideration of educational structures, both in schools and in the organization and focus of teacher training. Some would most likely argue that such a move would undermine the power and status attributed to mathematics.

Implications also reach into practices of researching mathematics education. Researching identities from a poststructuralist perspective is in itself an uncertain undertaking. First, a poststructuralist perspective renders it impossible to predict a priori identities and categories in research. Second, the dynamics and hybridity of identities and discourse prevent the establishment of fixed and objective results of research, in a traditional sense. Poststructuralist research, in its rejection of universalities, is also rejecting the hegemony of scientific results and the opinions of scientists. What research of this nature does offer is a reflexive interpretation of situated events and themes open to critique and alternative knowledge. It grants a voice to participants, acknowledging the locality of interaction. In such work, there is a distinct possibility for seeing, hearing, and learning about new aspects and characteristics of the construction of mathematics education.

The study presented and the extracts provided serve as examples of how participants in classroom interaction make use of identities and discursive practices in their interaction, which cannot be predicted by researchers, educational policy makers, or teachers, in their quest to enhance learning outcomes. The study shows how the mathematics classroom is not just for learning mathematics. It demonstrates how uncertainty is a continuous parameter to be reckoned with in the classroom interaction. Discourses and identities set the agenda and serve as possibilities for or obstacles to learning mathematics, rendering the learning process fragile. In the process, the study highlights the need to pay close attention not just to the cognitive processes of constructing mathematical knowledge, but also to the interaction surrounding these processes. Attention must be paid to everything that is not mathematical in the mathematics classroom.

NOTES

1. Skovsmose has been a continuous source of inspiration on the notion of un-
    certainty and its specific relations to mathematics education (see, for exam-
    ple, Skovsmose, 2005, 2006).
2. In research, the terms *postmodern* and *poststructuralist* are used with varying
    consistency, although there seems to be some agreement that the terms de-
    marcate a specific research perspective and place it in contrast to, for exam-
    ple, positivist research. In this paper, the term *poststructuralist* emphasizes the
    power of the uncertain and the fragile embedded in our being in the world
    and, therefore, also present in any attempt to carry out research. However, we
    draw inspiration from references claiming both a postmodern and poststruc-
    turalist perspective.
3. In Danish teacher training programs, it is mandatory to choose either Danish
    or mathematics as one of the specialized subjects. Choosing Danish can be
    difficult for students who do not have Danish as their first language.

REFERENCES

Alrø, H., & Skovsmose, O. (2002). *Dialogue and learning in mathematics education: In-
     tention, reflection, critique.* Dordrecht: Kluwer Academic Publishers.
Bauman, Z. (1996). From pilgrim to tourist–or a short history of identity. In S. Hall
     & P. Du Gay (Eds.), *Questions of cultural identity* (pp. 18–36). London: Sage.
Bauman, Z. (2004). *Identity.* Cambridge, UK: Polity Press.
Biesta, G. J. J. (2006). *Beyond Learning: Democratic education for a human future.* Boul-
     der: Paradigm Publishers.
Davis, P., & Williams, J. (2009). Hybridity of maths and peer talk: Crazy maths. In
     L. Black, H. Mendick, & Y. Solomon (Eds.), *Mathematical relationships in educa-
     tion: Identities and participation* (pp. 136–146). New York: Routledge.
Fuller, A. (2007). Critiquing theories of learning and communities of practice. In J.
     Hughes, N. Jewson, & L. Unwin (Eds.), *Communities of practice: Critical perspec-
     tives* (pp. 17–29). London: Routledge.
Hall, S. (1996). Introduction: Who needs 'identity'? In S. Hall & P. Du Gay (Eds.),
     Questions of cultural identity (pp. 1–17). London: Sage.
Kilgore, D. (2004). Toward a postmodern pedagogy. *New directions for adult and con-
     tinuing education, 102,* 45–53.
Kincheloe, J. L. (2003). *Teachers as researchers: Qualitative inquiry as a path to empower-
     ment.* New York: Falmer Press.
Klein, M. (2002). Teaching mathematics in/for new times: A poststructuralist analy-
     sis of the productive quality of the pedagogic process. *Educational Studies en
     Mathematics, 50,* 63–78.
Lange, T. (2007). *The notion of children's perspectives.* Paper presented at the The Fifth
     Congress of the European Society for Research in Mathematics Education,
     Larnaca, Cyprus (February 22–26).

Lerman, S. (2006). Cultural psychology, anthropology and sociology: The developing 'strong' social turn. In J. Maasz & W. Schloeglmann (Eds.), *New mathematics education research and practice* (pp. 171–188). Rotterdam: Sense Publishers.

Lyotard, J.-F. (1984). *The postmodern condition: A report on knowledge* (Trans: G. Bennington & B. Massumi). Manchester: Manchester University Press.

Sfard, A., & Prusak, A. (2005). Telling identities: In search of an analytic tool for investigating learning as a culturally shaped activity. *Educational Researcher, 34*(4), 14–22.

Skovsmose, O. (2005). *Travelling through education: Uncertainty, mathematics, responsibility.* Rotterdam: Sense Publishers.

Skovsmose, O. (2006). Research, practice, uncertainty and responsibility. *Journal of Mathematical Behavior, 25*(4), 267–284.

Stentoft, D., & Valero, P. (in press). Identities-in-action: Exploring the fragility of discourse and identity in learning mathematics. *Nordic Studies in Mathematics Education, 14.*

Tajfel, H. (Ed.). (1982). *Social identity and intergroup relations.* Cambridge: Cambridge University Press.

Turner, J. C., Hogg, M. A., Oakes, P. J., Reicher, S. D., & Wetherell, M. S. (1987). *Rediscovering the social group: A self categorization theory.* Oxford, UK: Basil Blackwell.

Valero, P. (2004). Postmodernism as an attitude of critique to dominant mathematics education research. In M. Walshaw (Ed.), *Mathematics education within the postmodern* (pp. 35–54). Greenwich, CT: Information Age.

Vithal, R., & Valero, P. (2003). Researching mathematics education in situations of social and political conflict. In A. Bishop, M. A. Clements, C. Keitel, J. Kilpatrick, & F. K. S. Leung (Eds.), *Second International Handbook of Mathematics Education* (pp. 545–592). Dordrecht: Kluwer Academic Publishers.

FURTHER READING

Kilgore, D. (2004). Toward a postmodern pedagoy. *New directions for adult and continuing education, 102*, 45–53.

Tajfel, H. (Ed.). (1982). *Social identity and intergroup relations.* Cambridge: Cambridge University Press.

Turner, J. C., Hogg, M. A., Oakes, P. J., Reicher, S. D., & Wetherell, M. S. (1987). *Rediscovering the social group: A self categorization theory.* Oxford, UK: Basil Blackwell.

CHAPTER 6

LEARNING TO TEACH

Powerful Practices at Work During the Practicum

Margaret Walshaw

ABSTRACT

Appelbaum (2008) tells us that "teaching is learning, and learning is teaching" (p. 1). I am interested in exploring how pre-service teachers come to teach and learn in the context of their mathematics practice in schools. My objective is to find out how certain versions of effective mathematics teaching, and not others, come to be intelligible to them. The focus is on the practicum experience, because it is in schools where possibilities and constraints of the pre-service teacher's understanding of herself as an effective teacher in the classroom are first fully confronted—where relationships are directly implicated and where multiple meanings are made. The practicum experience is where structural and organizational school processes and taken-for-granted understandings amongst practitioners shape the pre-service teacher's sense-of-self as a teacher in the classroom.

Successful practicum stories are often built around the cornerstone of community (Alton-Lee, 2003; Anthony & Walshaw, 2007). Community is about interactions between contexts and people—a *relation* between settings and

Unpacking Pedagogy: New Perspectives for Mathematics Classrooms, pages 109–128
Copyright © 2010 by Information Age Publishing
109

the people within those settings. Within the practicum, a sense of community develops from shared understandings of respective roles and an agreed upon meaning of pedagogical practice. Supervising teachers are key players in establishing the kind of community that will facilitate a productive practicum experience (Brown & Danaher, 2008; Darling-Hammond, Chung, & Frelow, 2002). Not only do they contribute to the professional learning of the pre-service teacher, they also, according to Sinclair (2008), influence "how," and, indeed, "if" the pre-service teacher's commitment to teaching will be sustained. Successful pre-service teachers, as Sinclair observes, are those who work within a professional community of shared knowledge of and shared thinking about pedagogical practice, and who are assisted both practically and emotionally through personal and systemic support.

Specifically, in this chapter, I am searching for insight about how teacher identity is negotiated within the structures and community of practitioners charged with the task of supporting pre-service teachers' development in schools. I am seeking to explain the central paradox of learning to teach, namely, that "there can be no learning without conflict, but the conflict that animates learning threatens to derail the precarious efforts of trying to learn" (Britzman, 2003, p. 3). Highly influential in the discussion have been Foucault's understandings of power and how subjects are produced and how intersubjective negotiations take shape. I draw upon Foucault's conceptual frame to understand how a group of pre-service teachers create a sense-of-self as teacher that is simultaneously present, prospective, and retrospective, as well as rational and otherwise.

INTRODUCING IDEAS

Foucault (e.g., 1984, 1988) provides a framework and a language for exploring the ways in which pre-service teachers develop their identities as teachers. His conceptual tools allow us to deal with the complex interplay between social structures and the processes of self-formation that are at work in learning to teach. For him, identity is always contingent and precarious (Walshaw, 2007). In this section, I introduce the concepts of discourse, power, governmentality, surveillance, normalization, and dividing practices, all of which are central to Foucault's conceptual toolkit.

Discourses

Discourses perform the role of conceptual schemes. They function like sets of rules, providing us with the knowledge about what is possible to speak and do at a given moment. They do that by systematically constituting specific versions of the social and natural worlds for us, all the while obscuring other possibilities from our vision. They are historically variable ways of specifying knowledge and truth. But they are more than that: The crucial point about discourses is that they position people in different ways as social subjects. In that sense, discursivity is not simply a way of organizing *what* people say and do; it is also a way of *organizing actual people and their systems*. For the pre-service teacher, this means that he or she *is* the production of the discursive practices through which she becomes subjected.

Power

Power is an overriding interest to Foucault. In *Discipline and Punish* (1977), he developed the themes of governmentality, surveillance, and normalization that allow us to explore processes of identification as they are lived by individuals in relation to both structural processes and lived experiences. According to Foucault, systems of power both produce and sustain the meanings that people make of themselves. This claim suggests that identities and subjectivities are strategically fashioned and contested through systems of power in the dynamics of everyday life. Foucault showed that power is distributed, rather than centralized; it operates from the bottom upwards, instead of from the top down; and it is positive and enabling. It is, however, also negative and coercive.

Governmentality

When Foucault talked of governmentality, he meant enforced obedience to rules that are presumed to be for the public good. It is the process by which our conduct is controlled in minute detail. A positioning as mathematics teacher is not governed by force or violence; rather, it is a more subtle coercion that keeps classroom practitioners in check. In his work, Foucault (1984) showed how the Panopticon—the Benthamite prison—in shifting its primary target from the body to the soul or self, "made it possible to substitute for force or other violent constraints the gentle efficiency of total surveillance . . . " (p. 217).

Panopticonism is barely visible in the classroom, yet the classroom is part of a wider educational regime whose central mission presupposes conforming and obedient individuals. Like other institutions, practicum schools operate as disciplinary technologies; they govern the work that both practitioners and pre-service teachers do. In a very real sense, practicum schools create the conditions for certain discourses, and not others, to be entertained, in relation to categories of being, acting, and thinking.

Surveillance

In Foucauldian terms, the practice of surveillance makes "it possible to qualify, to classify and to punish. It establishes over individuals a visibility through which one differentiates them and judges them" (Foucault, 1988, p. 184). Surveillance practices are important because they suggest practices of monitoring that are, simultaneously, practices for defining and potentially regulating subjectivities. For the pre-service teacher, particularities that relate to the supervising teacher, the classrooms, and the school at which the practicum takes place all have their place in constituting her as a "teacher." In particular, pre-service teachers are regulated through a covert set of standards and value systems associated with classroom behavior, thinking, speech, and actions.

In this way, the practicum school participates in practices that seek a hold on the pre-service teacher. The school produces mechanisms that will shape, monitor, and discipline the knowledges, modes of operating, and positionings of pre-service teachers for whom they are responsible. They create specific conditions and forms of control that will shape the pre-service teachers' behavior, their attitudes, and their pedagogical practice. They also construct particular positionings for them that both create and lend coherence to the understandings that pre-service teachers construct of themselves. That is to suggest that through subtle measures, the practicum school is extremely powerful in establishing the parameters around which pedagogical practice will be defined.

Foucault's understanding of surveillance is important because it suggests practices of regulation that are, simultaneously, practices for identity construction. Thus, it is impossible to discuss learning to teach in Foucault's terms without taking into account participation in the social practices of schooling. For the pre-service teacher, learning to teach is the initiation into a social tradition and into sets of "rules" operating at the practicum site. Regulatory practices at the site produce a certain network of material and embodied relations, controlling, classifying, and delimiting conduct in minute detail.

Normalization

Learning to teach is always relational: Pre-service teachers can only be teachers in relation to the meanings of others. The reason for this is that other people, as well as the systems that people have created, define what is normal. Through procedures that are made both explicit and implicit, schools regulate and exercise control over the meaning of teaching by normalizing and providing surveillance practices to keep such meanings in check. It is in that sense that, for Foucault, the notions of normalization and surveillance represent dual instruments of disciplinary power.

The pre-service teacher's self-understandings of what mathematics teaching should look like are mapped against certain criteria of effective teaching established within the school, determining the kinds of activity, behavior, speech, gestures, and networks that are legitimated. This set of criteria also endorses the way in which time and physical space will be controlled. When a pre-service teacher engages fully in the institutionalized practices of the school and of the supervisor's classroom within which she is working, she has learned to perform and enact the genres that constitute the knowledges, modes of operating, and theories and practices of the classroom. She also operationalizes the particular positionings and embodied practices that construct teaching in her supervisor's classroom.

It is possible to suggest that the pre-service teacher's engagement is "complete" when she is able to reproduce the particular identity constructed for her by her supervisor at her practicum school. Moreover, the pre-service teacher's engagement has become self-actualized when it does not appear to require monitoring and regulation. Using Foucault's language, the pre-service teacher actively involves herself in self-forming *subjectification*. By this is meant a positive process that involves the willing development and transformation of herself, involving disciplinary power, and with it, surveillance and normalization. She becomes accountable to the discourses that claim their hold of her in her supervisor's classroom, and she does this by disciplining/regulating herself without any apparent formal compulsion to do so.

In a very real sense, then, the practicum school performs the function of a technology of power, "determin[ing] the conduct of individuals and submit[ting] them to certain ends or domination, an objectivizing of the subject" (Foucault, 1988, p. 18). In turn, pre-service teachers within these institutions guarantee their compliance through "technologies of the self" that make it possible for them "to effect by their own means or with the help of others a certain number of operations on their own bodies and souls, thoughts, conduct, and ways of being, so as to transform themselves" (Foucault, 1988, p. 18).

Dividing Practices

Teachers, just like all of us, take up identities through processes of classification and division. Practices that Foucault calls *dividing practices* are fundamental to the way in which we differentiate teaching from other workplace practices. In all societies, and particularly in highly structured democracies, we make distinctions between people; and we make judgments based on the categories and differentiations that we have established. In schools, for example, students are objectified and classified according to intelligence testing, streaming/setting, performance on entry tests, records of achievement on school and national curriculum examinations, learning styles, participation and ability in sports and music, and so on. Knowledge about students' behavior, their achievements and the like, is developed through these dividing practices—they are instrumental in shaping the way we think about particular students. Students also come to think of themselves in ways that have been shaped for them, and begin to act accordingly.

Precisely because pre-service teachers are involved in a wide range of social practices, they will often be categorized quite differently from one context to another. Dividing practices that operate across social situations impact on them in ways that create a different sense of self. To this end, divisions operate not only between pre-service teachers, but also within individual pre-service teachers themselves. Dividing practices that are at odds with each other are most keenly felt by pre-service teachers as they move from one disciplinary institutional site to another. Each site attempts to regulate the pre-service teacher's behavior and her pedagogical practice.

APPLYING IDEAS

My exploration begins with Foucault's theory of language and social power, and takes seriously the discursive constitution of subjectivity. I examine moments of intersubjectivity in very localized sites and use this strategy to demonstrate how language "constructs the individual's subjectivity in ways that are socially specific" (Weedon, 1987, p. 21), and to reveal how discourses powerfully organize thinking and experience.

In applying Foucault's ideas to learning to teach, I draw on data from a study that took place on pre-service teachers' return to the university after their practicum during the third of four school terms[1]. These were second-year primary initial teacher education students, studying for their BEd. At the time of the research, each pre-service teacher had practiced and been appraised within five different schools during their university course. It was their fifth practicum experience that was the focus of this study. Although the earlier practica were not under interrogation, lessons learned from Foucault remind us that, in many subtle ways, the previous practicum experiences all played their part in constituting the pre-service teachers' sense-of-self as teachers.

Students in this course range in age from 19 to 53, and most are women. Only a small proportion is over the age of 40, and these are, in the main, change-of-career students. The lecturers from each of the classes offering the compulsory course worked as a team on the design, the schedule, the pedagogical approach, and the assessment of the course. I developed a questionnaire to open up a space for individual pre-service teachers to relate their practicum experiences and invited all internal students to participate in the study.

All 72 students in attendance on the data collecting day chose to participate. I asked them to respond to questions about their recent teaching practice experience. Through participation in the study, each pre-service teacher was given the opportunity to construct a sense of how teaching identity is created in relation to people and practices—how it is socially structured and historically inflected. Because I wanted to understand how pre-service teachers constituted *themselves* as teachers in mathematics and how they constituted *themselves* as moral subjects responsible for their own actions, the observations and reports of supervising teachers were extraneous to the analysis.

The investigation is grounded on the concrete details of the pre-service teachers' accounts, and care is taken not to gloss over their struggles during their time in schools. In doing that, I lay bare the complex ways in which practicum schools operate to regulate the work of pre-service teachers. I use quotes from what pre-service teachers write to provide "a profuse and diverse specificity . . . where voices are juxtaposed and counterposed so as to generate something beyond themselves" (Lather, 1991, p. 134).

Arguably, the questionnaire could only provide a partial and incomplete view of regulatory practices at work in learning to teach. However, the questionnaire provided information that led to new knowledge about processes and events in schools that contribute to the making of teachers.

Teaching Identity as Transitory

If, as in Foucault's understanding, teaching identities are constituted and negotiated within contexts of ongoing participation, then those identities are constantly on the move. Within and between different contexts, pre-service teacher identifications are marked by competing meanings of experience, circumscribed by differences in time, place, events, and commitment. It makes little sense, then, to reduce the complexity of pedagogical activity to a technical solution. At the beginning of their fifth practicum, pre-service teachers positioned themselves in relation to available discourses. During the practicum, other discourses came into play that in some ways confirmed and in other ways contradicted those earlier discourses. I explored ongoing transitional positionings by asking the pre-service teachers how they positioned themselves at the beginning and at the end of the practicum. In the following, I use words taken from their responses:

Beginning: I was looking forward to teaching. I felt confident and enthusiastic, though sometimes I am unsure about how to explain some maths ideas.

End: I was more confident and able to see students developing understanding.

Beginning: Prior to posting, I felt inadequate and concerned in algebra and problem solving.

End: I realize now that teaching any curriculum required me to learn the subject first. Teaching maths made me learn maths.

Beginning: I felt a bit uncomfortable—wanted more in-depth knowledge of how to teach it. How to get it across. How to start and introduce a unit.

End: More confident. Timing and pacing is now more sorted out. I learned which resources, activities, and strategies worked for students.

Beginning: I was lacking confidence.

End: I knew the material well and I was comfortable with maths. However, I still lacked management strategies.

Beginning: I felt that I was familiar with the resources and activities, but I had no idea how to run the maths lessons.

End: I was still very in the dark with what she did with each group individually. That's because I was always helping the very new group.

Beginning: Scared of teaching maths.

End: Feeling a lot more relaxed. Children really seemed to enjoy the activities, especially using money.

Beginning: I was a bit shaky and nervous about teaching maths, as it is not my favorite subject. I feel it is necessary to give children a sound base of understanding in mathematics and mathematics skills, so I was therefore skeptical of my own ability.

End: I appreciated being able to teach at different levels and gain strategies for teaching each level. I feel a lot more confident in teaching maths. However, I am going to further my maths education by taking maths papers [electives] next year.

Neither wholly "student" nor "teacher" in the classroom, the pre-service teacher's identity during the practicum is mapped onto a complex grid of formal and informal educational discourses and practices. They brought to the practicum their notions of mathematics teaching constructed through familiarity with the pedagogical relation as a student in schools. They also brought their personal student experiences in the university degree course and their experiences at former practicum schools. Steeped initially in the as-yet-still-developing self-constructions as teachers, many were aware that they entered a new range of discourses and identities that would constitute them as a teacher of mathematics. On the threshold of something new at the beginning of the practicum, slightly over fifty percent of the pre-service teachers worried about the development of their identity as a teacher and how they might constitute the teaching of mathematics within their own subjective experience of mathematics.

During the fifth practicum, new aspects of the teacher's world were laid bare, and new relationships with teachers, administrators, and students were made possible (Britzman, 2003). This new context operated to define its own borders, representing different and competing relations of power, knowledge, dependency, commitment, and negotiation. It was within the new context that pre-service teachers identified and named categories of unknown knowledge, namely, *the teacher's mathematical explanations; algebra and problem solving; starting and introducing a unit; how to get it across; how to teach it; how to run the maths lessons; general mathematical knowledge.* By naming these categories, pre-service teachers had established their own personal classificatory grid for the development of a teaching identity for elementary mathematics. At the end of the practicum, most pre-service teachers registered a growing confidence. The categories they identified became less general and more specifically tied to the construction of actual pedagogical practice within the classroom. Variously named as *timing, pacing, resources, activities, workable strategies, management strategies; teaching to groups; teaching different levels; intention to enroll in an elective maths paper*—all these knowings became the coordinates through which "good" teaching would be mapped.

Technologies of Normalization

Pre-service teachers are not only redefining their teaching identities in relation to available discourses in the classroom and to the complex selves of others, they are also learning what is defined as "normal" practice through the school's organizational processes, routines, and timetabling. Granted, institutional practices are measures and techniques that produce identities implicitly, rather than by repressive force, yet the rationalities underpinning their specific ways of doing and knowing have the same purpose of regulating individuals. The forms of subjectivity the practicum school valorizes are sanctioned in accordance with how it organizes and strategizes for time and space. Such regulatory practices are constructed for the normalization and, ultimately, the production of the self-governing individual teacher. In that sense, not only is the practicum school a site of production and regulation of teachers' subjectivities, it is also the site for the regulation of pedagogical practice. In that the organizational practices within institutions function as techniques that normalize teaching practice and produce specific teaching identities (Walshaw, 2007), they are extremely power.

To address the question of how principles of organization, as developed by schools, are implicated in the production of governable individual pre-service teaching, I asked pre-service teachers to state the "Usual time of day for mathematics lessons," "Usual duration of lessons," and "Total number of mathematics lessons occurring during teaching practice."

The pre-service teachers in this study stated that mathematics was routinely taught in the morning (89%). For fifty-six percent, mathematics was scheduled for early morning, and for a third, mathematics took place between the morning break and lunchtime. Timetabling arrangements like these that differentiate curriculum areas impose temporal conditions through which the pre-service teacher is obligated to teach a designated curriculum area and discouraged from teaching any other. Arguably, the widespread practice of morning mathematics, taken together with commonsense understandings of positive effects of morning learning, would suggest that the social significance of mathematics is not lost on schools.

A third (30%) of the pre-service teachers saw mathematics taught on a daily basis, and only a relatively small number (15%) reported that mathematics took place less than four times per week. Pre-service teachers quickly learned how long each lesson would be programmed, given that the scheduled length of time was consistent from one day to the next. However, the expected duration varied considerably from one classroom to another. While the median time spent on mathematics during the school week was three hours and twelve minutes, one pre-service teacher came to expect five hours regularly each week. Practices of administration like these induce the pre-service teacher into a particular cyclic order, in which specific tasks and functions, by turn, are to be performed. And because these institutional practices fix limits, controlling the "time" around which pedagogical reality might take place, they foster the development of and regulate what is to count as mathematics teaching.

Within the classroom itself, the supervising teacher authorizes particularities that regulate minute details of space and time. Those particularities also regulate bodies, reaching into the most intimate private thoughts and desires, to the extent that they produce and normalize movements and observable bodily practices. Two pre-service teachers made the following comments about their supervising teachers: "It was amazing to watch her in action. She definitely loves to teach, and maths has a very high profile in her class." "My Associate was the leader of the school maths program, and her knowledge and love for maths definitely helped me." For these two pre-service teachers, the supervisor's work in the classroom operated through individual movements and gestures, all the while creating the differential relation effective/ineffective teacher. The production of competency here worked through very situated and continuous micro-practices of power, in the most seemingly trivial details of embodied practice. It is in this way that systems and regimes of effective teaching are produced and reproduced. As Foucault (1984) notes, "[w]hat makes power hold good, what makes it accepted, is simply the fact that it doesn't weigh on us as a force that says no, but that it traverses and produces things, it induces pleasure, forms knowledge" (p. 61). Spoken and embodied discursive relations can be read as an artifact of the professional collaboration achieved within these two supervising/pre-service teacher relations.

If the practicum school is a site of production and regulation of teachers' subjectivities and a site for the regulation of teaching, so, too, is the university course a powerful factor in the construction of an identity as a teacher. By advancing an ideological construction of a mathematics teacher and by producing mechanisms to shape the knowledges, modes of operating, and positionings of pre-service teachers, the course created the conditions for certain discourses, and not others, about what it means to be a teacher of elementary school mathematics, to be recognized. Foucault puts it this way: "Every educational system is a political means of maintaining or modifying the appropriateness of discourses with the knowledge and power they bring with them" (Foucault, 1972, p. 46).

The university course work, on the one hand, imposed specific categories of being, acting, and thinking about what effective mathematics teaching is like. Supervising teachers in schools, on the other hand, invest in their own particular discursive codes of mathematics pedagogy, which foreground particular processes and practices for planning and enacting practices in the classroom. In asking pre-service teachers to describe a typical lesson, I attempted to understand the part that the university course work plays in framing their classroom observations. I wanted to understand the ways in which the course work functions as part of the technology of normalization.

Like many educational practices, the typical lesson structures pedagogical arrangements for school work, establishing a set of practices and social relations for the teacher and learner in the classroom. According to pre-service teachers' observations, in most classrooms, the teacher first *maintains prior knowledge, introduces new concepts for the day, making links with prior knowledge, provides explanations, models, poses questions for the children, supplies work and activities to enable practice of these mathematical ideas*; and, finally, *reflects on the work*. In this logic, the teacher moves reflexively from talk, to writing on the board, to observing, to talk and questioning, all the while grounding understanding through the process of children's activity and written work.

However, each classroom produces its own truths about teaching practice, and what is taken as "true" in one classroom is not to be considered as universal, nor, indeed, even necessary in others. Each classroom operates with an established set of rules of formation, through a network of material and embodied relations that invest the individual with certain aptitudes and predispositions toward a particular identity. What is important is that the process of shaping and informing positionings operates surreptitiously to facilitate the production of a normal, conforming individual. Teaching practice in any one classroom becomes intelligible through its reliance on certain strategies that are accepted, sanctioned, and made to function as true. Power, knowledge, and truth become central to the constitution of relationships in the classroom.

> Each society has its regime of truth, its "general politics" of truth: that is, the types of discourse which it accepts and makes function as true; the mechanisms and instances which enable one to distinguish true and false statements, the means by which each is sanctioned; the technique and procedures accorded value in the acquisition of truth; the status of those who are charged with saying what counts as true. (Foucault, 1984, p. 73)

Each classroom has its particular *regime of truth*, which legitimizes and sanctions a discursive space for certain practices and social arrangements. Pre-service teachers observed, in most classes, teacher talk and exposition, and children engaging in whole discussion and debate and working with hands-on equipment. Children often worked on worksheets. Pre-service teachers estimated that group and cooperative activities were assigned for half of the class time. A quarter of the pre-service teachers observed peer assessment and a third of the children marked and corrected their own work. Theoretical decisions about learning like these have important implications for the ways in which pedagogical relations can be conceptualized and enacted. In creating particular modes of activity, ways of being, and interpersonal relationships, such decision making makes possible both what can be said and what can be done within the elementary mathematics classroom. Knowledge, then, including practitioners' knowledge, is implicated not only in the practices of administration, but also in the production of forms of sociality.

Transitioning from the university course to the practicum school moves the pre-service student towards a different network of political and social discursive practices. New discourses come into play, and the identity positions and politics that these discourses offer provide them with access to a differential engagement and positioning in relation to the regime of "knowledgeable" practice operating as effective mathematics teaching. These discursive codes and how they are to be taken up are not always consistent with previous experiences, nor even consistent across pre-service teachers. Nor are they always made explicit to the novice: "Maths just seems to happen in this classroom. It just arrives along like all the other curriculum areas" (respondent). To the supervising teacher, however, teaching constitutes a tight script that establishes how the teacher's work is to be enacted in the classroom. Not only are pre-service teachers, then, working at embodying the technicalities of practice and behavior for their supervisor's classroom, they are also, among other things, exploring their positionings in relation to histories of standards and value systems of the supervising teacher.

In order to explore the "take up" by pre-service teachers of their supervisors' practices, I asked them to comment on the way their expectations were met regarding the way in which mathematics was taught. Eighteen percent chose not to answer this question. Forty-seven percent claimed that their expectations were not met. Some of those responses follow:

> Worksheets weren't great all the time. I expected to see more hands-on work as well.

> No use of equipment.

> Disappointed that only maths text books and worksheets were used.

> I felt that by not having group work that some children were slipping through the gaps.

> I didn't like children working through a textbook—this was not my idea of teaching.

> I think the Associate [supervisor] sees herself as a facilitator to pass out worksheets.

> I almost felt as if I knew more about teaching maths and portraying it effectively.

However, located in other classrooms with other supervisors, thirty-five percent of the pre-service teachers stated that the way in which mathematics was taught had met their expectations:

> HANDS ON! Children did enjoy the practical activities.

> Group work went well, as children are closer in ability.

> [I was interested to see] that concepts were put into real, relevant contexts and that children were able to experience these.

> The teacher integrated maths into the morning roll call, as children counted how many children were at school, how many boys/girls, the difference between number of boys/girls, etc.

Teacher always asked "How did you work that out?" and got children to explain their working out.

Math was very much made relevant and hands-on for the children, who experienced a lot of different activities, e.g., popcorn (mass/weight), cooking recipes, different food containers.

Of interest was the way many children in my class supported each other in their work, or were willing to tutor each other.

The university course work, through the knowledges and modes of operating that it advocated and promoted, had established a baseline understanding of "teaching mathematics." Through explicit engagements with the official curriculum statement and its theoretical representations of development, cognition, pedagogy, assessment, and the learner, pre-service teachers had learned what counted as evidence of effective teaching practice. They knew what particular pedagogic modes were legitimated in the course and the types of classroom arrangements deemed central to knowledge facilitation.

In this study, the "privileged teaching repertoire" (Ensor, 2001, p. 299) of the university course promoted the use of *apparatus and technology*, it recognized *difference*, and it validated *problem solving, group activities, integrated learning*, and *collaboration*. The teacher's role was to create a supportive learning environment, facilitating and empowering, rather than posturing as the authoritative knower. Thus, the teacher was expected to create a productive learning community and to provide tasks conducive to the construction of mathematical knowledge (Anthony & Walshaw, 2007). She was to promote whole class discussion and small group collaborations, in which learners' explanations and justifications were given opportunity for expression. She, as teacher, was also expected to know when to intervene and when not to interfere (Walshaw & Anthony, 2008). Thus, through a covert set of standards and value systems associated with mathematics teaching at the elementary level, a construction of the effective teacher was naturalized and made inevitable.

[Power] ... applies itself to immediate everyday life which categorizes the individual, marks him by his own individuality, attaches him to his own identity, imposes a law of truth on him, which he must recognize and which others have to recognize in him. It is a form of power which makes individuals subjects. (Foucault, in Dreyfus and Rabinow, 1982, p. 212)

Technologies of Surveillance

Crucial to the formation of an identity as a teacher in the supervisor's classroom, then, are the surveillance practices that monitor and sustain meanings of teaching through subtle and diffuse ways. A successful practicum in the classroom derives from a power that circulates through the pre-service/supervisor relation.

> My associate was very well organized and supportive. She shared all her plans and resources with me. She provided quality feedback with positive ideas for me to improve on.

> She [the associate] gave me a lot of freedom to use my ideas. She supported me and asked if I needed anything and shared resources.

> Full of ideas, very supportive of new approaches. Happy to share information. Associate happy to learn herself.

Not only are pre-service teachers working to understand the "performance" of mathematics practice, they are also, among other things, exploring their teaching positionings in someone else's classroom. Lacking the full credentials to live in the world of teaching, pre-service teachers attempt to establish a teaching identity in a classroom that is already populated with others' views. Their work, focusing on embodying those practices that will elevate their subordinate and less influential position, is placed under the panoptical gaze (Foucault, 1977), and assessed against the supervisor's standards. It is those very practices of surveillance—regulating and sanctioning the work of the pre-service teacher at the classroom level—that provide continuities within the pedagogical site.

That is to say that in the supervisor/pre-service teacher relation, the pre-service teacher is one of the primary effects of disciplinary institutional power, the most pervasive disciplinary practice being, to borrow Foucault's term, "the gaze." "The gaze" is delicate and seemingly intangible, yet its networks are invasive, seeking to construct what teaching will be like in a particular classroom. It seeps through bodies, gestures, and behavior. For the pre-service teachers quoted above, subtleties within the networks of power were shaping a love of and passion for teaching mathematics. Such alliances, however, were not always apparent:

> [At the end,] I was disillusioned at the lack of encouragement I received from my A. T. [Associate Teacher]. Nice enough person, but I think that I made her feel I was taking her class away, as they were so responsive to all the new ideas that I brought into the classroom. I could have used some help in developing new ways for the children to think and try things, but the A. T. had tried and true methods of working, and that was the way it was.

> After slowing down, I got the hang of taking a maths lesson. I tried doing more exciting activities with the class, which they enjoyed, but I found after a couple of days it was best for them to go back to the structured routine of book work...The lessons should not be structured so much so that children can't handle change.

> I was enthusiastic and ready to put ideas to practice about how to be an effective teacher, but the topic-based maths programme didn't allow for it.

> I wanted to introduce new ideas, but did not have enough confidence. I just followed my Associate's plans. I felt I could not try new things, as my Associate was set in the way things were done.

> At first, I was very enthusiastic and full of ideas, but found that, due to the teaching style of my associate, it was difficult to implement my plans. [At the end,] I had adjusted my personal style to fit the class culture. It is difficult to force your way. You really just have to fit the class as it already is.

> I was forced to follow her methods of teaching in maths, as that is what she had planned and wanted maintained. I am confident in maths, but was given little opportunity to express my confidence. Could not go outside the square.

Differences, deviations, peculiarities, and eccentricities of practice come under interrogation. In the view of the supervisors, these pre-service teachers demonstrated both a limited and a limiting discourse of classroom life. In these classrooms, processes of surveillance worked surreptitiously to equalize behavior, actions, and even thinking, in the most seemingly innocuous details of embodied practice. Highlighting difference and marginalizing the idiosyncratic, the supervisor naturalized particular constructions and excluded all others that did not comply. Like the pre-service teachers in Britzman's (2003) study, the practicum experience highlighted differential institutional practices, contradictory realities, and competing perspectives. For these pre-service teachers, the new space was fraught with tensions, and resulted in ambiguous and sometimes painful negotiations to produce an individual identity as a teacher in the classroom.

Learning to teach is a strategic and interested activity, dependent on embodied relations of power. Teaching practice in schools always works through vested interests, both those of the practitioner and those of the pre-service teacher. Those vested interests are not simply to do with the present: The past, as well as anticipated experiences also play their part in the construction of a teaching identity and in the designation of effective teaching.

For Foucault, the pre-service teacher who invests fully in the discursive practices of the supervisor is ever mindful of others. Against this Foucauldian reading, theories of individualisation (e.g., Beck, 1992; Giddens, 1994) claim that it is the subject who is central, and is of his or her own making. Such theories claim that, motivated by a desire for fulfillment of their own interests, individuals are responsible for negotiating their own destiny. The individual, it is argued, "must render his or her life meaningful, as if it were the outcome of individual choices made in the furtherance of a biographical project of self-realization" (Rose, quoted in Walkerdine, 2003, p. 240), irrespective of the constraints encountered. But for many pre-service teachers working in schools, this is a simplistic reading of the situation: others always get in the way. The range of choices is clearly not as extensive for some pre-service teachers as Beck's thesis of individualization would lead us to believe. Indeed, powerful discourses worked to constrain the identities that some pre-service teachers desired for themselves. For them, networks of power operating within the supervisor and pre-service relation deeply implicated the way in which they practiced in the classroom.

Learning to teach is the continuing task of authoring oneself. It represents an ongoing struggle to resolve the contradiction between one's current positioning with diminished resources and less social power, and the discursive place where one might command one's own destiny. Becoming a practitioner requires systems knowledge: figuring out what you can and cannot have, working out what is open and what is not open for negotiation, and then determining the remaining options and their likely costs.

CONCLUSION

This chapter has examined how social and structural processes interact in the shaping of those learning to teach. Applying insights from Foucault's scholarly work, I have explored the ways in which the realities of the social and material structures in schools play out for those learning to teach. The view was toward developing a sensitivity of the impact of regulatory practices on pre-service teachers' constructions of themselves as teachers. Through their reflections of learning to teach in schools, it was possible to develop an understanding of how teaching identity is produced and reproduced through social interaction, daily negotiations, and within particular contexts, which are already overburdened with the meanings of others. In revealing the way in which particular practices are normalized and regulated within the practicum experience, the political and strategic nature of teaching practice came to the fore.

What a Foucauldian interrogation allows is an examination of how social relations and structures that are often considered to be relatively independent of power influence the development of practice and identities. It helps explain the "simultaneous articulations of a dispersed and localized shifting nexus of social power" (Haywood & Mac an Ghaill, 1997, p. 268) surrounding those learning to teach in schools. Explanations that refocus the learning-to-teach story away from personal narratives toward the operation of networks of power have profound implications for mathematics education. Learning to teach becomes much more than narrativization. Theories that hold individuals accountable for their own fate tend to overlook the part that significant others play in authoring one's own biography in the practicum context. As Thomson, Henderson, and Holland (2003) argue, such theories "underplay the importance of relationships and forms of reciprocity and obligation that are embedded within them for understanding the identities and practices in which individuals engage" (p. 44).

In the Foucauldian understanding, learning to teach becomes an issue of micro-political engagement with discursive classroom codes, all of which are set upon providing the pre-service teacher with a sense of identity in the classroom as a teacher. Becoming a teacher is not so much an issue of a personal journey as a barely visible set of highly coercive practices. Teaching "know-how," then, is linked to networks of power, targeting thinking, speech, and actions, with a view toward producing particular constructions of identity. It is the result of compliance to a set of practices that have been naturalized for the pre-service teacher in the classroom. Such an understanding points to the "reiterative power of discourse to produce the phenomena that it regulates and constrains" (Butler, 1993, p. 2). If there is a freedom of choice, it is a freedom constrained by "a lineage of loose alliances, relations of resistance and mastery, and configurations of fluid interests... [that are] not outside the games of truth" (Blake et al., 1998, p. 62).

NOTE

1. An earlier version that reports on this study appeared in the *Journal of Mathematics Teacher Education, 7*(1).

REFERENCES

Alton-Lee, A. (2003). *Quality teaching for diverse students in schooling: Best evidence synthesis.* Wellington: Ministry of Education.

Anthony, G., & Walshaw, M. (2007). *Effective pedagogy in mathematics/pāngarau: Best evidence synthesis iteration [BES].* Wellington: Learning Media.

Appelbaum, P. (2008). *Engaging mathematics: On becoming a teacher and changing with mathematics.* New York: Routledge.

Beck, U. (1992). *Risk society: Towards a new modernity.* London: Sage.

Blake, N., Smeyers, P., Smith, R., & Standish, P. (1998). *Thinking again: Education after postmodernism.* Westport: Bergin & Garvey.

Britzman, D. (2003). *Practice makes practice: A critical study of learning to teach* (2nd ed.). Albany: New York Press.

Brown, A., & Danaher, P. (2008). Towards collaborative professional learning in the first year early childhood teacher education practicum: Issues in negotiating the multiple interests of stakeholder feedback. *Asia-Pacific Journal of Teacher Education, 36*(2), 147–161.

Butler, J. (1993). *Bodies that matter.* New York: Routledge.

Darling-Hammond, L., Chung, R., & Frelow, F. (2002). Variation in teacher preparation: How well do different pathways prepare teachers to teach? *Journal of Teacher Education, 53*(4), 286–302.

Dreyfus, H., & Rabinow, P. (1982). *Michel Foucault: Beyond structuralism and hermeneutics.* Sussex: The Harvester Press.

Ensor, P. (2001). Taking the 'form' rather than the 'substance': Initial mathematics teacher education and beginning teaching. In M. Van den Heuvel-Panhuizen (Ed.), *Proceedings of the 25th Conference of the International Group for the Psychology of Mathematics Education* (Vol. 2, pp. 393–400). The Netherlands: Freudenthal Institute, Utrecht University.

Foucault, M. (1972). *The archaeology of knowledge and the discourse of language* (Trans: A. Sheridan Smith). New York: Pantheon.

Foucault, M. (1977). *Discipline and punish: The birth of the prison* (Trans: A. Sheridan). Harmondsworth: Penguin Books.

Foucault, M. (1984). *The care of the self* (Trans: R. Hurley, 1986). Harmondsworth: Penguin.

Foucault, M. (1988). *The ethic of care for the self as a practice of freedom.* Cambridge: The Massachusetts Institute of Technology Press.

Giddens, A. (1994). *Beyond left and right.* Oxford: Polity Press.

Haywood, C., & Mac an Ghaill, M. (1997). Materialism and deconstructivism: Education and the epistemology of identity. *Cambridge Journal of Education, 27*(2), 261–272.

Lather, P. (1991). *Getting smart: Feminist research and pedagogy with/in the postmodern.* New York: Routledge.

Sinclair, C. (2008). Initial and changing student teacher motivation and commitment to teaching. *Asia-Pacific Journal of Teacher Education, 36*(2), 79–104.

Thomson, R., Henderson, S., & Holland, J. (2003). Making the most of what you've got? Resources, values and inequalities in young women's transitions to adulthood. *Educational Review, 55,* 33–46.

Walkerdine, V. (2003). Reclassifying upward mobility: Femininity and the neo-liberal subject. *Gender and Education, 15*(3), 237–248.

Walshaw, M. (2007). *Working with Foucault in education.* Rotterdam: Sense Publishers.

Walshaw, M., & Anthony, G. (2008). The role of pedagogy in classroom discourse: A review of recent research into mathematics. *Review of Educational Research, 78*(3), 516–551.

Weedon, C. (1987). *Feminist practice and poststructuralist theory.* Cambridge, MA: Blackwell.

FURTHER READING

Foucault, M. (1970). *The order of things: An archaeology of the human sciences* (Trans: A. Sheridan). New York: Vintage.

Foucault, M. (1972). *The archaeology of knowledge and the discourse of language* (Trans: A. Sheridan Smith). London: Pantheon.

McNay, L. (1994). *Foucault: A critical introduction.* Cambridge, MA: Polity Press.

Walshaw, M. (2007). *Working with Foucault in education.* Rotterdam: Sense Publishers.

CHAPTER 7

REGULATING MATHEMATICS CLASSROOM DISCOURSE

Text, Context, and Intertextuality

Elizabeth de Freitas

ABSTRACT

Mathematics teachers perform through discursive patterns of instruction and curriculum, while also positioning (often unconsciously) themselves and their students in relation to various cultural constructs and contexts. Teacher discourse in the classroom often contains comments *about* mathematics as discipline and mathematics as text, and such comments frame and contextualize the activity within the classroom, and reveal teacher assumptions regarding the ontological and epistemological status of mathematics texts. I use the word *text* to refer to all forms of semiotic and linguistic interaction, but, in this chapter, focus primarily on written and spoken texts. Teacher references to the relationship between these texts and "real" mathematics or the "real" world function as framing devices in classroom interaction, whereby a "context" for classroom activity is construed. Not only do such discursive habits constitute an arguably false binary between the mathematics beyond the surface of the text and the text itself, but they also position students and teachers in power relations structured and sustained by the leveraging of

Unpacking Pedagogy: New Perspectives for Mathematics Classrooms, pages 129–151
Copyright © 2010 by Information Age Publishing

this binary. As a *discursive formation* within classroom discourse, such meta-statements about mathematics texts regulate and govern student and teacher engagement with the very texts that are, as Veel (1999) suggests, the source of "differentiated access to meaning potential" (p. 206). Thus, the discursive construal of the relationship between text and context plays a pivotal role in regulating student access to meaningful mathematics.

This complex discursive performance points to the multiple ways in which teachers frame the ontological status of texts, while simultaneously positioning themselves and students in power relations with respect to "mathematical knowledge." Positioning is not simply the act of situating someone within particular discursive practices; positioning is an inevitable and dual process whereby subjects are subjected to dominant normative discourse, while also constituted and enabled through such compliance (Butler, 1997). Our discursive habits involve a contradictory mix of *submission* to the cultural norms inscribed in the discourse, and *empowerment* through this act of re-inscription. Pedagogy always involves this contradictory mix of submission and empowerment, since teacher and student voice and authority are constituted by the very discourse that confines and disciplines their actions. Teacher and student agency thus emerges through this contradictory mix; they are hailed as subjects within an institutional discourse. In mathematics classrooms, in particular, teacher agency is uniquely constructed in relation to highly esoteric texts, and teacher authority is frequently enacted in terms of decoding and contextualizing these texts, be it within the discipline itself, the "real" world, or prior habits of textual production.

In this chapter, I try to trouble the taken-for-granted distinction between text and context by examining classroom transcripts for evidence of how mathematics texts are discursively framed by teachers. My aim is to show how contexts are realized in discourse, and that we therefore have no unmediated access to them. Thus, I advocate for a poststructural approach to the notion of "context" as a "hypertextual pastiche" (Lather, 2007, p. 110), an approach that lays the groundwork for reading subjectivity through discursive patterns of interaction. In arguing for a textual reading of context, and in attempting to disrupt the often unexamined binary between text and context, I hope to make room for a decoding strategy that reveals the positioning of subjectivity through particular grammatical features of classroom discourse. I use the concept of intertextuality (Lemke 2002; Tannen, 2007) as a means of theorizing the fusion of text and context in classroom interaction, and argue that patterns of citation, repetition and reference to 'other' texts create a textual layering which functions to produce and position subjects within classroom discourse.

INTRODUCING IDEAS

Critical Discourse Analysis

Critical discourse analysis assumes that utterances in the classroom occur in relation to various orders of discourse. Orders of discourse are entrenched normative habits of social interaction and are central to processes of cultural and material reproduction (Fairclough, 2003). Equitable access to the cultural capital of school mathematics depends on teacher and student acculturation into these orders of discourse. An order of discourse both confines what is possible in any given interaction and also produces the space for such possibility. For instance, repeated teacher references to the *difficulty* of a particular mathematics text can function as regulative discursive moves that position students and teachers in relation to cultural norms regarding ability and achievement. Students are then labeled as having "math difficulties" when the meaning of such texts remains un-accessed. Critical discourse analysis aims to describe the structure of these interaction patterns, and to examine the shape and form of linguistic mediation in relation to these orders of discourse. Within this framework, language use is assumed to be a form of action that produces meaning and positions speakers in particular power relations. A focus on the intersection of language use and power relations is crucial in understanding how students and teachers are constituted through classroom discourse. Fairclough (2003) argues that subjectivity, authority, and agency are regulated by orders of discourse that are mapped onto the "linguistic features" of texts (p. 29). Power relations are "linguistically realized" (Fairclough, 1999, p. 189) in particular lexico-grammatical features of texts, which thereby signify specific socially ratified forms of knowledge and experience.

In this chapter, I grapple with a methodological issue that haunts all studies of classroom discourse, namely, the challenge of decoding discourse so as to understand the way that *subjectivity* is constituted and enacted (sometimes in contradictory ways) in brief and seemingly spontaneous utterances. In other words, I am grappling with the challenge of mapping the socio-cultural features of subjectivity onto the lexico-grammatical features of particular utterances. A 2005 survey of work completed under the banner of critical discourse analysis asked that researchers address this challenge and better articulate the links between "the grammatical resources and the social practices" under discussion (Rogers et al., 2005, p. 386). The challenge emerges most strongly in studies in which critical discourse analysis is employed as a means of decoding the power relations of classroom interaction. In the case of mathematics classroom discourse, this challenge is compounded by the apparent depersonalized language. Rogers et al. (2005) criticize critical discourse analysis for being too ambitious and for frequently reading ideology into the data, and suggest that Halliday's systemic functional linguistics is more cautious in differentiating between ideological inscription and other functions of language. But analyses based only on systemic functional linguistics have "tended to lose sight of people in their focus on texts" (Tannen, 2007, p. 12). My aim, in contrast, is to center questions of subjectivity and agency, to simultaneously attend to grammatical features of texts, and to thereby trouble the assumed divide between text and context.

Mathematics texts are often assumed to be neutral and their meaning independent of context. One can argue, however, that mathematical texts, like other kinds of language use, are a form of action that position speakers in particular power relations, and thereby constitute and structure the specific context where meaning emerges (Atweh, Bleicher, & Cooper, 1998; Christensen, Stentoft & Valero, 2008; de Freitas, 2008a; Morgan, 2005; Walshaw, 2004; Zevenbergen, 2003). This argument seems particularly relevant to studies of mathematics classrooms where texts are "recontextualized" (Bernstein, 2000) and knowledge is *re-produced* for specific pedagogical purposes. Linguists such as Veel (1999), O'Halloran (2005), and Christie and Martin (2007) have effectively applied systemic functional linguistics and the sociological theories of Bernstein to school mathematics texts, arguing that the "instructional discourse" is embedded in "a discourse of social order (regulative discourse)" (Bernstein, 2000, p. 49). Christie and Martin (2007) focus on the distinctive linguistic features that make "the language of mathematics different from the language of other school disciplines" (Veel, 1999, p. 186). Indeed, the technical lexicon, the hierarchical ordering of concepts, and the esoteric grammar of school mathematics constitute what might be considered a "monstrous" language—a language that refuses everyday meanings. According to Martin (2007), this monstrous "vertical discourse" resists "common sense" primarily because of the excessive use of nominalization, by which processes and actions are reconfigured as nouns. For instance, algebraic equations such as $3x - 11 = 15$, are formally translated into dense noun phrases like "the difference between three times a number and 11 is 15," whereas students tend to verbalize these equations in terms of temporal actions and operations, such as, "take a number and multiply it by 3 and then subtract 11 to make 15." This process of reifying or "thingifying" (Christie & Martin, 2007) the dynamic process of acting and operating (that is to say, construing mathematical processes as noun phrases) dominates the written form of mathematics discourse, and reflects the set-theoretic foundations of the field, while being entirely at odds with student and teacher verbal interaction. Martin (2007) suggests that the extreme nominalization in the vertical discourse of written mathematics is possibly what makes it such a gate-keeping discipline. Indeed, the radical disjunction between the written and the spoken in mathematics classrooms, through nominalization and other linguistic and semiotic practices, functions to construe spoken discourse as a framing discourse that positions teachers as mediators of written texts.

Text, Context, and Intertextuality

It seems reasonable to claim that we are always interested in language use *in context*, and yet we have radically different understandings of what that context might include (and what it might exclude). According to Halliday (2007), the original meaning of the word *context* was, in fact, opposite to its common usage today, in that it actually referred to the text that accompanied the object of discussion (and hence, it was the *con*-text). Only during the twentieth century did the term come to refer to the non-verbal "setting" in which language is used. This setting was variously interpreted to mean the way that language as a system was situated within a "culture," and the way that instances of language use occurred within a "situation" (Halliday, p. 271). These two different ways of interpreting setting, drawn from different linguistic traditions, point to different ways of theorizing language itself; in the first, language is a system that models or reflects culture, while in the second, language is an enactment of situational social relations. Halliday draws on both to help found his systemic functional theory of language use:

> What the learner has to do is to construe (that is, construct in the mind) a linguistic *system*. That is what is meant by "language as a system": it is language as stored up energy. It is a language, or some specific aspect of a language, like the language of science, in the form of a *potential*, a resource that you draw on in reading and speaking and listening—and a resource that you use for learning with. How do you construe this potential, and how do you use it when you've got it? You build it up, and you act it out, in the form of *text*. "Text" refers to all the *instances* of language that you listen to and read. And that you produce yourself in speaking and in writing. (Halliday, 2007, p. 274, italics in original)

Halliday is careful to point out that the "system" of language is not some fixed ideal structure, but rather "the potential that lies behind all the various instances" (2007, p. 275). Thus, the act of "instantiation" is not the selection of one element from a finite set of choices (since the meaning potential of all texts is, in fact, infinite), but, instead, a mediated response to the demands of the "situation." Moreover, the relation between a situation and a text, where the situation is "realized" in the text, is not *causal*, but rather "semiotic" (Halliday, 2007, p. 282). This crucial point—that the relationship is semiotic and not causal—underscores the fact that context and text are mutually construed and that the relationship of instantiation or realization must be seen as bi-directional (Halliday, 1985, 1993).

The distinction between text and context appeals to our common sense, but it tends to conjure an unexamined ontological divide, and thereby disrupts all attempts to read subjectivity through text, and text through subjectivity. Nor does the distinction between semiotic and causal assist when addressing the power relations operating through language. Halliday sets in motion a rethinking of the semiotic relationship between text and context, but poststructural approaches to subjectivity have gone further in problematizing the distinction, revealing how there is no getting outside of text, where text is used as a tag for semiotic mediation of all kinds (Derrida, 1974). Thus, there is no pre-textual context, no subject outside of text, and no pure text disengaged from other texts. This theoretical development is crucial for recognizing the complex disciplinary power of textual production and for avoiding the inequity that ensues from essentializing the "real," in terms of the text/context binary.

Within this poststructural tradition, critical discourse analysis conceives of "context" in terms of the multiple textual framings within a text. In other words, the context is construed by a sequence of overlapping and nested textual references through which various meanings are produced, which are, themselves, part of the texturing of discourse. Each text construes and constructs a context through these references to other texts. Framing the context of the interaction involves explicit and implicit reference to these other texts, be they past statements or written documents, or, perhaps most importantly, social practices that are evoked as normative. Framing can be thought of as a means of constructing social borders around texts, and thereby defining an "outside" to the text. Briggs (1993) argues that the issue of context requires researchers to expand their close textual readings of transcripts with a sustained attention to these references to other texts within the text. He focuses on how context is performed through this layering of textuality and how it functions as a "means of creating, sustaining, and/or challenging power relations" (p. 391). In this chapter, I will use the term *con/text* as a means of both referencing the history of the term, as well as underscoring its poststructural revisioning.

Although the term *intertextuality* comes from Bakhtin ([1952], 1986) and Kristeva (1980), and one finds in their work a theory of "speech genres" in relation to subjectivity, it is the linguists who have developed this notion as an analytic tool (Tannen, 2007). Intertextuality is primarily about repetition, collocation, and pattern within discourse, insofar as texts refer to other texts already in circulation. Discursive events are, thus, always intertextual, insofar as all forms of interaction involve recitation, representation, and reference to other texts. This reworking of prior language use shapes the current con/text and positions subjects within the discourse. Tannen (2007) states: "'[I] ntertextuality,' then, in its many guises, refers to the insight that meaning in language results from a complex of relationships linking items within a discourse and linking current to prior instances of language" (p. 9). Tannen refers to the work of the linguist Becker (1995), who offers a set of six categories for identifying forms of intertextuality. These six are helpful in grasping the diverse forms of linking that occur in interaction, and they point to the way that texts point to other texts, while emphasizing that language use always involves this relational linking at multiple levels. Becker's categories are: (1) structural relations (between part and whole), (2) generic relations (of texts to prior texts), (3) medial relations (of text to medium), (4) interpersonal relations (of text to participants in a text-act), (5) referential relations (of text to nature and to "the world one believes to lie beyond language"), and (6) silential relations (of text to the unsaid or the unsayable) (cited in Tannen, 2007, p. 11). In the next section, I show how these various intertextual relations are manifest in mathematics classroom transcripts.

According to Fairclough (1993), the intertextual approach recognizes the inherent heterogeneity within particular orders of discourse (regarding style or truth claims), while at the same time studying the way these different orders leverage different discursive resources and produce material power relations. Intertextual analysis shows how texts selectively draw upon different orders of discourse in ways that serve to validate particular positions within the discourse. By shifting our attention back and forth between the instantiations of *texts* in the classroom and the *orders of discourse* that are invoked in each text, we generate a more textured socio-cultural reading of classrooms, and are able to point to the power relations enacted in all forms of interaction. Lemke (2002) argues that the practice of constructing intertextual relations "does the social work of reconstituting the social relations of the community" (p. 39) and that "language functions ideologically not only by enabling us to make meanings that maintain the dominance of one group over another but also by not providing ready-made ways of challenging that dominance" (p. 40). Tracing the "linguistic realization" of orders of discourse involves a double-reading of texts, so that both the work of *analysis* and *interpretation* are achieved. In other words, mapping correlations between the instructional and regulative discourse demands a reading at the close micro-level and, simultaneously, at the hermeneutic macro-level. For instance, the teacher's re-voicing of student utterances in more formal mathematical language is often part of an interaction that involves the building of consensus, and which can be studied to learn how teachers delicately manage and marshal consent. The grammar of these utterances can be analyzed in relation to the dominant discourse of control and consent.

APPLYING IDEAS

Positioning the Subject

The data under discussion emerged from a research project designed to explore the complexities of mathematics teacher identity. Four mathematics teachers were studied in their classrooms during one semester. All four taught in the same rural Canadian high school with an economically diverse, but primarily white, student population of about 700. Notes and transcripts from the class observations were analyzed using a discourse analysis framework. Observation notes from all of the twenty observed classes recorded teacher-directed pedagogy (whole class instruction and teacher transmission) and an overwhelming emphasis on mastery of procedure (in contrast to inquiry instruction or conceptual investigations). The data were first coded for shifts between three kinds of orders of discourse: mathematical (inquiry, procedural, conceptual), administrative (assessment, management, school issues), and con/textual (personal narrative, anecdote, metaphoric framing, application). These were then subdivided into eight registers: (1) procedural, (2) conceptual, (3) inquiry questions, (4) personal narrative, (5) anecdotal, (6) metaphoric, (7) classroom management, (8) school business. Note that the original categories of "mathematical discourse" and "con/textual discourse" were dispersed over 8 registers, some of which used resources from both. The transcripts were analyzed for occurrences of and transitions between these registers.

A register enacts rules or conventionalized practices of language use. The different registers I used for transcript analysis are usually found in classroom discourse, and, in some cases, are blended or mixed. A register involves more than a specialized vocabulary set. Each register has its own social grammar, and each thereby interpellates a distinct subject or identity. Both the speaker and listener are constituted through the form of address. The meaning of an utterance is constructed within the social grammar and the implied power relations of the register. In order to comprehend an utterance, the learner and speaker must locate the utterance within a particular register, and, simultaneously, the learner and speaker must also recognize themselves as constituted as subjects through the register. For instance, when speaker and listener use a narrative register about personal experience, they are hailing each other as personal acquaintances. The personal narrative register often enacts power relations of intimacy, exposure, and vulnerability. The procedural register, on the other hand, is characterized by rigorous rule-following and the imperative mode, and thereby constitutes facets of subjectivity that are radically different. In the procedural register, both speaker and listener are addressed in terms of: (1) proficiency, (2) compliance, (3) abstraction/absence, and (4) atemporality.

In this section, I discuss the two male mathematics teachers, Mark and Roy, from the larger study, who taught, respectively, grade 11 and grade 12 university preparatory courses. I focus first on Mark, who framed each mathematics lesson by beginning and sometimes closing with questions from what he called the "life quiz." Students were given two minutes to ruminate on questions such as, "Would you leave your country and never return for a million dollars?" Or "Who has it easier, women or men?" followed by a brief discussion about their answers. Students were also responsible for having researched geographic and historical information about a country—each assigned one at the beginning of the year—and were told to be prepared to share three minutes of their research at Mark's request. Mark used these non-mathematical topics to engage students at the outset of class, and to create whole-class conversations, in which many students engaged. He also embedded many references to sports throughout his lessons, and he had an ongoing conversation with five boys about the national hockey league results. The sports conversations often erupted, as though unbidden, during the mundane procedural discourse that dominated his classroom.

Mark repeatedly used metaphors to construe an antagonistic relationship between students and texts in that "a lot of the math that I'm going to throw at you" came at the students aggressively. He used a Rottweiler rating system to announce the difficulty of each problem written on the blackboard. Variations of "This is a puppy Rottweiler" were used to signify relatively easy problems, and "This is a hungry Rottweiler tied to a post" to signify relatively difficult problems. Near the end of the lesson on the complex number system, he said, "We're in a Rottweiler cage right now, being bit at the heels."

The lesson under discussion in this chapter opened with ten minutes of "life quiz" questions and a two-minute student description of the Czech Republic. Mark then shared a personal narrative about traveling in the Czech Republic, where his passport was taken from him at the border at gunpoint, and he worried that he would be imprisoned. What follows the dramatic narrative of his near imprisonment in the Czech Republic—his use of the personal narrative register—is a radical shift to the mathematical register, using a framing device that demotes talk of personal narrative and personal preference as the "aside" text. His explicit use of these con/textual openers for his lessons functioned to underscore the de/con/texualizing of the mathematics register. In this particular lesson, Mark introduces the complex number system, and further frames the topic in four distinct ways: (1) through repeated references to previous mathematics texts and habits of engagement with those texts, (2) through reference to the "real" world and its influence on the historical development of mathematics, (3) through the construal of "mathematics" as a habit of rule construction for ease and practicality in thinking and doing, and (4) through grammatical forms of address that position students in terms of their facility at decoding mathematics texts.

Mark: Okay. Anyway, just a little aside. (with reference to the dramatic personal narrative)

> **Mark:** Okay, in your notes, under May 18, we are going to stretch your mind. (writing the date on board)
>
> **Mark:** Somebody asked me the other day when we were doing quadratic formulas and they said, you know, when am I ever going to use this in real life, and I said, you won't, okay? 1.1% of us might use it in "real life," but what you will use is the ability to figure something new out. Okay? The other day, they asked me for a job reference, because they're going to try and get a new job at Subway. Now, they're going to get into Subway and they're going to start working, and they're going to have to figure out how to use the new cash register. There aren't cash registers where you just press a button; you've got to figure out the computer, touch the screen. You've got to figure these things out. Okay. So, what a lot of the math that I'm going to throw at you, is about figuring something out new, a foreign idea. It doesn't have to do with practical applications so much, okay? Most of you know how to feed yourselves, dress yourselves, you know, wash yourself pretty well. Okay? So that sort of stuff is done. What we have to do now is be able to throw different scenarios at you and see if your brain can handle it. Now, today is a perfect example of that, because this is mathematics that is beyond my application, beyond your application, but it's the understanding of a completely foreign idea, okay.

Mark's reference to the complex number system as a "foreign idea" is a tacit bridging to the previous discussion about negotiating foreign border authorities. The antagonistic relation he experienced with these authorities is mirrored in the construal of mathematics coming "at" the students, as though they were under attack, and also fuels his moral imperative to rise to the occasion and negotiate "different scenarios." Shifting registers is a complex linguistic capacity, and it often involves the introduction of framing devices to assist in the re-positioning of both the speaker and the listener. When a speaker moves abruptly from one form of address to another, he or she often enacts two radically different and sometimes mutually incomprehensible forms of subjectivity. The shift involves more than two specialized sets of vocabulary in two different discourses. Speakers are subjected by the generic structure of the register and are made compliant to the rules of the discourse. It is often at the point of suturing two registers together when the delicate work of re-positioning the subject occurs; framing devices are like linguistic markers to assist listeners in recognizing themselves in the coming discourse. It is also through the meshing of distinct registers that metaphors emerge. When two different register codes are juxtaposed and integrated in a meaning-making process, the texts are re/con/textualized in relation to each other, and they mark each other, and deposit a trace. These traces—like "foreign idea"—are generated both consciously and subconsciously. Not only is this trace the site of metaphor production, but the suturing of two registers also reveals the framing devices that operate to position speakers in relation to the two codes, be they oppositional or other. Mark shifts registers by framing mathematics as an activity that will ultimately prove useful to students as they problem solve in the "real" world. The radical cultural divide between the complex number system and a cash register speaks to Mark's rhetorical skills at framing, but, ironically, it also operates tacitly to validate the extreme disjunction between text and con/text in Mark's classroom. It also underscores various orders of discourse circulating in the classroom, in particular, the "myth of utility" (Dowling, 1998), whereby mathematics is inscribed as a master narrative into and onto all other practices.

After the initial framing of mathematics in terms of a foreign idea "beyond your application," Mark then walked the students through a sequence of simple equations, as though recounting the history of their learning trajectory in terms of exposure to new kinds of numbers. He reminded them that, over the years, they had repeatedly been told to revise their notions of what constituted a possible solution to a given equation, tracing their exposure to the natural numbers, the integers, the rational numbers, and the irrational numbers. The story of student encounters with these kinds of number systems was couched in terms of both "we" statements about habits of activity and "I" statements about Mark (or another teacher) as the authority who stipulated whether an equation had a solution or not. Then Mark made an abrupt shift and very briefly described Leonard Euler in the 1700s, mentioning alternating current and how "the amplitude of that actually is measured with something that has to do with the square root of negative one." Given that he had yet to identify the taboo equation that was now to be granted a legitimate solution, and that no one had yet heard nor considered the problem of the square root of negative one, this con/textual historical framing seemed somewhat out of place, but I believe it functioned, like many other register shifts in Mark's discourse, as a highly ritualized regulative discursive formation that actually afforded him the possibility to de/con/textualize the mathematics texts (for more, see de Freitas, 2008b). He then turned abruptly and wrote on the board $x^2 = -9$ and said:

> **Mark:** I've told you until this point that you can't take the square
> root of negative numbers, and in the real number system,
> we know that the square root of 9 is 3, but when you came
> to something like that [points to $x^2 = -9$ on board], you said,

uh, can't do it. . . . Well, they discovered that, actually, it has an application for the sign of the waves in electrical currents and in light waves. Particles that are going through light that we see actually have a wave. They can measure it, and it has to be used using what we call imaginary or complex numbers. So, we're going to give you, I'm going to give you just a little touch of what these numbers are all about. Now, what I would like you to write down is this. They had to come up with a new set of numbers, and it's all based on one idea. [writing, while speaking] "i" is the square root of negative one. Okay. They took that, and they said okay, "i" is going to stand for imaginary numbers [writes "imaginary numbers" on board]. Somebody said it was sort of an unfortunate name, because we think imaginary numbers all look like, you know, the Easter bunny or something, but actually these things are real, it just tends to not be something that is very easy to understand, for most of us, including me. I find it hard to understand. [pause] So, imaginary numbers are based on this idea that i is equal to the square root of negative one. Now, remember when we were doing our radicals, and we wanted to break down the root of 50? How did we break it down?

It is amazing how quickly Mark abandons his con/textual motivation and shifts to the procedural mastery of manipulating the new symbols. One can see how rapidly a grammatical metaphor (the introduction of i as a marker for the "real" process of wave propagation) takes shape, and how quickly the history of its emergence is erased. This brief con/textualizing discourse is followed by 30 minutes of procedural mastery. His "little touch of what these numbers are all about" is quickly supplanted by the citation of previous habits in "Now, remember when we were doing our radicals . . . ," and the rest of the lesson is devoted to mastery with increasingly more complex symbolic expressions.

The use of pronouns in this excerpt is revealing, in that it positions Mark, the students, and the mathematics in terms of different kinds of authority. He opens the register with "I've told you until this point that you can't . . . ," which immediately posits his "social authority" (Solomon, 2009, p. 183) and the student submission to that authority via the high polar modality of "you can't. . . . " He then switches to "we" when returning to the recounting of their previous exposure to the real number system. This "we" functions as a marker or token of a comfort zone, where one and all are included. But suddenly he shifts pronouns to "you" in "when you came to something like that . . . ," which disrupts the shared comfort of prior knowledge and positions the students in isolation, precisely when he wants to isolate them so as to dramatically draw their attention to the unknown. At this point of isolation or separateness, Mark introduces the "they" in "Well, they discovered that it actually has an application . . . ," which validates the authority of mathematics as a disembodied "social" authority outside the current textual production. Thus, in the moment of their isolated vulnerability, the students are subjected to an external social authority that stipulates the legitimacy of the text. This last con/textual authority ("they") is actually mystified through the talk about alternating current, insofar as Mark fails to offer an adequately detailed explanation of the connection to the "real," and yet the students know not to ask for more details regarding these ritual exercises in con/textualizing.

Mark contrasts the "real" of imaginary numbers with the not-real of imagined Easter bunnies. This framing of mathematics as radically different from benevolent phantom creatures is only partially a concern that the students may misinterpret the vocabulary; it is also a coded message to the students that they remain in an oppositional relation to mathematics texts. Indeed, the contrast is drawn so as to make clear that mathematics remains something "that is not very easy to understand" and that the texts they will encounter will resist them.

The Difficulty of Calculus

I focus now on the chair of the mathematics department, Roy, who had been teaching for over thirty years. He taught the upper level grade 12 academic classes, and was positioned as the gatekeeper and benign authority who would shepherd students on to university. He was observed on five occasions, and in each case, his lesson began with his asking students to identify homework questions they were unable to complete, some of which he then completed on the blackboard, before introducing a new "topic," and proceeding to treat a series of problems from the textbook, and finally assigning homework from the textbook and allowing students to commence the homework in class. Roy enacted the role of the expert and veteran teacher. Observation notes indicated that procedural mastery was extremely valued, and that students were motivated to succeed on formal assessments. I have chosen to focus on a set of textual citations that seem to interrupt the dominant procedural register of symbolic reproduction. These appear as interruptions because they are not always coherently integrated into the dominant register; and yet—importantly—they function to underscore the legitimacy of the procedural register. Teacher references to prior and future tests, for instance, often suddenly emerge in the midst of a long algebraic derivation on the blackboard, and thereby frame the current production of a text, as Roy decides to remind the students that questions like the one under consideration will be seen again. Many of these interruptions are meant to manage student expectations concerning the difficulty of the task, and also to make explicit teacher expectations about student capacity to perform in the future. The difficulty of calculus was repeatedly mentioned during the five observations. Part of the implicit social contract between Roy and the students was that he would diffuse this difficulty.

The focus on difficulty is extremely important because of the way it directs Roy's attention to the negotiation of texts, rather than the negotiation of meaning. Roy works with the surface features of the texts, assuming that doing so will make them more accessible. Focus on difficulty also causes Roy to constantly step outside of the current textual production and reflect on the grammatical features that must be mastered in order to succeed. Talk of difficulty thus maintains the procedural register and directs student attention to patterns in the grammar of the texts, while also texturing the conversation with another layer of reference—a layer that construes power relations between speakers—as Roy repositions himself as the mediator between difficult texts and student learning. The discourse of difficulty is thus a regulating order of discourse that functions to maintain the emphasis on procedure, while positioning the speakers in relation to the authority of both the teacher and the mathematical register. In the excerpt below, Roy moderates the student perception of the difficulty of optimization problems, not entirely diffusing their anxiety, but offering assurances that their familiarity with the procedure for handling these kinds of problems will serve them well when they see them again on the test.

> **Roy:** Right. Like volume is always cubed. That's a tough problem, right?
> **S1:** Will a problem like this be on the test?
> **Roy:** Will it be on the test? Well, Patty, you know me and sine and cose. [laughter] You say I have this thing for sine and cose. Well. Will this really be as tough the next time around?
> **S2:** Yeah. Probably.
> **Roy:** Now you know how you handle it.
> **S2:** Ugh. If you're giving us a problem with a sphere or whatever...
> **Roy:** I'll give you the volume.

Roy closes this exchange with the promise that he will fulfill his part of the deal and supply the students with the required information—in this case, the volume. The continual references to difficulty are intertextual and function as (1) warnings on how to engage the text, (2) support in the face of a text resistant to understanding, and (3) subsequent evaluation of the skills required for interpreting the text. These continual interruptions function to mark Roy as the authority who can decode the symbolic text, but rather than decode the text in terms of meaning, he decodes it in terms of difficulty:

> **Roy:** If you spend your time on these, people, the next test won't be too bad. I'm not going to throw a lot of different situations at you. If you've done the work, you get the reward. Boy, number 6 is the same situation as 3a, only asking a different situation.

When Roy says "I'm not going to throw a lot of different situations at you," he uses the same expression "throw" that Mark uses, and, like Mark, he positions himself as benign authority, someone who has the power to subject the students to something far worse, but who has chosen to reveal what future mathematical texts will entail, and in what way these future texts will mimic the current texts. This promise that future mathematical texts will be like the ones currently produced implicates the teacher as the controlling agent, and situates his language use within the regulative discourse.

His use of the word "situation" is significant in this case, because he is discussing optimization problems, all of which have some con/textual application written into them, and often cite a "real life" context, such as maximizing the profit of a business venture or the volume of a box. One can see the delicate weave that joins Becker's two categories of intertextuality in this instance—the interpersonal and the referential—as Roy's utterances about future applications position speakers in relation to a textualized physical world, in that the situations to which he refers are entirely scripted and coded by the symbolic practices of the procedural discourse. The con/text of the application is entirely illusory—rather, construed as such—and the students are assured that the teacher has the authority to author the construction of these coded physical worlds. Indeed, the textualizing of these applications is crucial, in terms of Roy's pedagogical objectives. He has to assure the students that applications are essentially textual exercises. Like other high school calculus courses in the region, optimization problems and related rates of change problems comprise the two main areas of application, and are considered the most challenging by many teachers. In promising that the situations will be similar, he is erasing the contingent factors of the con/texts elicited in each problem, and informing the students that they can also consider these situations as coded and scripted, and within the procedural register. This message—that the application problems are texts to be decoded—is a way of controlling their meaning.

> **Roy:** Well, here's what you can take to the bank. Of these questions, if I ask you for a minimum, what you're going to find is a minimum. Okay?

Again, we can see the benign authority that Roy performs, assuring the students that his word is good (as good as gold?), in that the procedure for finding extrema will produce exactly what *he* asked for. He needs to call on this authority because of the ambiguity built into the procedure—that being that it might generate maxima and minima, and that con/textual features need to be analyzed in order to determine which is generated. Roy positions himself in relation to the power of the mathematical register, claiming an authority to marshal that power in directions determined by his own instructions. One can see how such a statement erases all significance from the con/textual features of the application, and further entrenches authority in his evaluation of the difficulty of the given text. When a student who had not been present at an earlier class continues to ask about the difference between maximima and minima, Roy elaborates below. I have italicized the personal pronouns to draw the reader's attention to them:

> **Roy:** Oh, *you* weren't here. Right. Not with respect to the way *we* do these problems. I know *you* know the difference between maximum and minimum, but when it comes to doing these problems, no, *you* do them both the same way. *You* can trust that if I ask for a maximum, what *you*'ll find is a maximum, not a minimum. And *you* might say, "What are *you* talking about?" Yeah. How do *we* find maximum/minimums? If *we* have a function, take the derivative and set it equal to zero. Now, remember, that identifies maximums and minimums. And *you* might say, "How do *you* know which one *you* really found?" I'd say, if *you* really want to prove that, *you* sub it into the second derivative and *you*'ll find out if it's a maximum or a minimum. But I'm also saying that if I ask *you* for the greatest, what *you*'ll find will be the . . .

Roy pauses deliberately at this point, and a student completes Roy's final sentence with "greatest," thereby indicating to Roy that he understands his end of the bargain in the power conditional statement. The student's response—"greatest"—is a discursive act of submission, whereby he demonstrates his capacity to anticipate the teacher's words and fill in the teacher's blanks. But there is more to this than simply the pattern of initiate–response–evaluate, in which teacher–student interaction consists of this sort of compliance. A closer analysis of the grammar of this exchange reveals how Roy is performing an imaginary dialogue between himself and the students, a dialogue that reconfigures student confusion as student interrogation of the meaning of the procedure. When Roy says, "And you might say, 'What are you talking about?'" and again, "And you might say, 'How do you know which one you really found?'" he is simulating a dialogue in which the teacher must defend the meanings behind the procedure. Roy's enactment of an imagined dialogue between himself and his students, using this technique of imagined student voice, reconfigures classroom discourse as debate about the meaning of that discourse. This discursive move is another example of intertextuality, specifically, a form of a *silential relation* or fictive citation, in that the conversation to which he refers never actually occurred, and probably could not occur, given the discursive norms in his class. Roy is offering a meta-discourse that mimics an ideal classroom interaction in which various speakers have equal power to contribute. He is acting out an imagined classroom discourse that hasn't happened. This is an example of intertextual production, whereby meaning is introduced and monitored through reference to the textual production ("you might say…") within the interaction. In doing so, he creates a space where he can regulate the debate and control any interpretations of the text. Roy used this technique of simulated dialogue on a number of occasions when he needed to resolve student confusion.

Taking a closer look at the grammar of this exchange reveals more along these lines. His use of the second person pronoun *you* functions in multiple and interesting ways as a means of positioning himself and the students in relation to each other, and in relation to the mathematical register. In this brief exchange, he uses *you* to signify (1) the actual student who had been absent, (2) the imagined student who might demand an explanation, (3) the collective students who must follow his instructions, and even (4) himself as the one who is addressed by the imagined student demanding an explanation. He quickly shifts from addressing the one student in "you weren't here" to addressing the entire class. His next use of the word *you* in "And you might say…" positions the class as the site from which a question demanding meaning might emerge, and Roy models that critique by asking the question on their behalf. In using "might have," he opens up a space for them to consider future interactions in which they actually *do* ask these questions. The use of "might have" is an example of modality, whereby a speaker chooses a particular measure of certainty or confidence to express meaning. Roy chooses the low modality of "might," which allows him to revoice one student's question as though it were the question spoken by all students, since "you might have" is, indeed, addressed to the entire class. Once the move is made to address all possible students, the "might have" begins to take on more of a "should have," with stronger modality. This possible decoding is supported by Roy's ongoing investment in his authority in the classroom, and reflects his didactic approach.

The next use of *you* functions to disrupt Roy's authority and reposition him as "other" to the classroom discourse, since he is confronted (in the imaginary dialogue) with the accusation that his words fail to make sense ("What are you talking about?"), and hence, his authority to speak is in question. Indeed, "What are you talking about?" is a fascinating choice of words for a number of reasons: First, it is unlikely that a student would choose these words in Roy's class, as there is a certain rudeness to them that would not be accommodated, and second, it is the first reference to a substantial *what*—meaning something non-discursive—that might lie behind the text. For both reasons, the statement is another marker of the silential intertextuality in Roy's classroom. The question stands out in the transcript as a demand for a con/textual reference, a reference to something beyond the textual production of procedure. And yet, the fact that Roy produces this question in an imaginary student voice, with a somewhat higher pitch than his own, diffuses its power and the power of the student who might consider asking it.

You then functions quite differently in the next imagined student question ("How do you know which one you really found?"), since this *you* is now the anonymous collective *you* of the plural other (the masses) who follow procedure without understanding. This latter *you* is tied to the action of the procedure, and the act of procedural submission, whereas the former *you* is the *you* that interrogates meaning: "What are you talking about?" can be decoded as "Which meaning are you signifying in this case?" while "How do you know which one you really found?" can be decoded as "After we follow your instructions, how do we know which text to produce?" Decoding these sentences allows us to map the discursive authority in this brief interaction.

These different meanings and forms of address constitute different subject positions within the discourse: the very first *you* declares the prior absence of the student; the second *you* tentatively constitutes a subject position that empowers students to demand meaning behind the procedures; the third *you* constitutes a subject position for Roy whereby he is the producer of meaning (the speaker who determines the *what* to which he refers); and finally, the last *you* constitutes a subject position with minimal agency—the agency of procedural enactment without understanding. Roy and the students are multiply constituted through this simulated dialogue—this text within a text.

The transcript continues with Roy proceeding to complete the second derivative for an example on the board, and to then evaluate the second derivative at the point in question:

> **Roy:** Now, if I put 40 in here, what kind of number do I get? Positive or negative?

> **Student:** Positive.
>
> **Roy:** Positive. And if you get a positive, it means you are working with a local . . . minimum. Alright. Positive for minimums, negative for maximums. So there's the proof that what we found is the minimum. Hard on the head?

Note that the "proof" is merely a procedural tool—developed in previous classes—to be implemented in cases when the results from other procedures are ambiguous. Roy's commitment to helping students succeed repeatedly causes him to reduce processes to grammatical metaphors through nominalization, as in "positive for minimums, negative for maximums." This indicates that teacher privileging of procedure and the consequent investment in the "monstrous" language of mathematics may, ironically, reflect teacher commitment to increased access to and facility with the symbolic register. The added question, "Hard on the head?" is meant to convince the students of the simplicity of the situation. Roy is asking them to assess the difficulty of the task themselves, or rather, to assess their own ability to solve the problem. "Hard on the head" is an interesting metaphor because of its embedded mixed messages regarding student learning. On the one hand, it can be interpreted as "Are you hard of hearing?" indicating once again Roy's emphasis on submission or capitulation to the order of discourse behind the text, or it could be interpreted as "Is your head made of wood?" indicating Roy's belief that student ability (or intrinsic lack of ability) needs to be explicitly addressed as a form of negative motivation. Or it could be interpreted as a verbal "banging over the head," by which Roy uses the mathematical register as a metaphor for physical punishment. All of these interpretations are plausible, given that the dominant regulative order of discourse in Roy's classroom pertains to the difficulty of calculus, and the comment addresses the students in terms of ease and comfort (in addition to ability), and thereby constitutes the "good student" as the one who has no difficulty in completing these tasks.

CONCLUSION

Fairclough (2003) expands on Halliday's theory of grammatical instantiation and selection from a system of "meaning potential," and extends it to study the ways that particular linguistic registers are framed within larger orders of discourse. Multiple layers of choice operate when texts are produced, indicating how texts are, in fact, forms of action that re-inscribe the legitimacy of certain discursive practices *and* particular orders of discourse, all the while positioning subjects in relation to each other. Choice also involves the exclusion of orders of discourse, in that specific texts are produced by *not* selecting and validating other texts.

The advantage of such an approach to studying mathematics classrooms is its emphasis on the possibility of choosing differently, and the consequent potential to enhance inclusion, but the problem with such an approach is that discursive choices are primarily unconscious, and speakers rarely reflect on the repertoire of other choices they might have made. The term *choice* should not be associated with enlightenment notions of an unbound individual *will*, free to determine its path. Speakers use language without reflecting on all of the alternate phrasing and possible permutations they might have used instead. They may frequently deliberate over using one word over another, but they do not choose their subject positions—as though they were hats or boots—and they rarely see language use as an act of submission (or resistance) to the power relations produced through the interaction. Language use is highly unreflective, and becomes increasingly so when one considers the social meanings embedded deep within the grammars of our registers. Attention to the complex enactment of diverse discursive formations, however, can help focus our attention on the orders of discourse inscribed onto the seemingly neutral texts circulating in classrooms. Moreover, through the analysis of language use, one can begin to see the contingency of these subject positions and power relations, and begin to imagine otherwise.

Critical discourse analysis is ambitious in its aim to correlate linguistic practices with forms of subjectivity construed and positioned through those same practices. Multiple levels of analysis and interpretation are required to subtly braid together the examination of lexico-grammatical features of texts with socio-cultural features of orders of discourse. Doing so is particularly challenging in studies of mathematics classroom discourse because of the assumption that the meanings of mathematics texts are independent of con/text and position of speaker. In this chapter, I have argued that the binary between text and con/text must be disrupted through a poststructural reading, so as to override assumptions about the neutrality of mathematics texts. My hope is that the concept of intertextuality functions as a tool for tracing the textual layering of references within a text by which subjectivity and the "outside" are construed and organized. In the con/text of the classroom, intertextuality can be used effectively to map the framing of student and teacher subjectivity. A focus on intertextuality allows researchers to study the discursive patterns of reference by which an "outside" of the classroom is construed. Patterns of reference to other texts are precisely what constructs a con/text for the given interaction. The material conditions of student and teacher school experiences are no less significant—indeed, this approach is meant to better honor those material conditions by recognizing how they are multiply constituted and entrenched through discursive practices in the classroom.

By focusing on the framing devices found in Roy's and Mark's classrooms, I have tried to map a correlation between the grammatical patterns of textual production and the regulative discourse that structures the power relations between teachers and their students. Attention to the grammatical details of these forms of address and the subsequent texturing of classroom discourse can shed light on the way students and teachers mediate mathematics texts. My hope is that this combination of analysis and interpretation will help expose the discursive mechanisms of symbolic domination whereby some students are positioned "outside" the text of classroom discourse. In Mark's case, the con/texts he used as framing devices were ritual exercises in authorizing the procedural discourse. In Roy's case, despite his good intentions to mediate the difficulty of calculus, his repeated reference to difficulty functioned to further rarefy the mathematical register and make it all the more "monstrous."

REFERENCES

Atweh, B., Bleicher, R. E., & Cooper, T. J. (1998). The construction of the social context of mathematics classrooms: A sociolinguistic analysis. *Journal for Research in Mathematics Education, 29*(1). 63–82.

Bakhtin, M. M. (1986). The problem of speech genres (Trans: V. W.McGee). In C. Emerson & M. Holquist (Eds.) *Speech genres and other late essays* (pp. 60–102). Austin: The University of Texas Press. (Original work published 1952).

Becker, A. L. (1995). *Beyond translation: Essays toward a modern philology*. Ann Arbor: University of Michigan Press.

Bernstein, B. (2000). *Pedagogy, symbolic control and identity: Theory, research, critique* (revised ed.). London: Taylor and Francis.

Briggs, C. L. (1993). Metadiscursive practices and scholarly authority in folkloristics. *Journal of American Folklore, 106*(422), 387–434.

Butler, J. (1997). *The psychic life of power: Theories in subjection*. Stanford, CA: Stanford University Press.

Christensen, O. R., Stentoft, D., & Valero, P. (2008). Power distributions in the network of mathematics education practices. In E. de Freitas & K. Nolan (Eds.) *Opening the research text: Critical insights and in(ter)ventions into mathematics* (pp. 147–154). New York: Springer Verlag.

Christie, F., & Martin, J. R. (Eds.) (2007). *Language, knowledge and pedagogy: Functional linguistic and sociological perspectives*. Continuum.

de Freitas, E. (2008a). Troubling teacher identity: Preparing mathematics teachers to teach for diversity. *Teaching Education, 19*(1), 43–55.

de Freitas, E. (2008b). Enacting identity through narrative: interrupting the procedural discourse in mathematics classrooms. In T. Brown (Ed.), *The psychology of mathematics education: A psychoanalytic displacement* (pp. 139–155). Rotter-dam: Sense Publishing.

Derrida, J. (1974). *Of grammatology.* Johns Hopkins University Press.

Dowling, P. (1998). *The sociology of mathematics education: Mathematical myths/ pedagogic texts.* London: The Falmer Press.

Fairclough, N. (1999). Linguistic and intertextual analysis within discourse analysis. In A. Jaworski & N. Coupland (Eds.), *The discourse reader* (pp. 183–212). New York: Routledge.

Fairclough, N. (2003). *Analyzing discourse: Textual analysis for social research.* New York: Routledge.

Halliday, M. A. K. (1985). *An introduction to functional grammar.* Victoria, Australia: Edward Arnold.

Halliday, M. A. K. (1993). Towards a language-based theory of learning. *Linguistics and Education, 5*, 93–116.

Halliday, M. A. K. (2007). The notion of "context" in language education. In J. J. Webster (Ed.), *M.A.K. Halliday: Language and education.* London: Continuum Press. (Original work published 1991).

Kristeva, J. (1980). *Desire in language: A semiotic approach to literature and art* (Trans: T. Gora, A. Jardine, & I. S. Roudiez). I. S. Roudiez (Ed.). New York: Columbia Press.

Lather, P. (2007). *Getting lost: Feminist efforts toward a double(d) science.* Albany, NY: SUNY Press.

Lemke, J. (2002). Ideology, intertextuality and the communication of science. In P. H. Fries, M. Cummings, D. Lockwood, W. Spruiell (Eds.), *Relations and functions within and around language* (pp. 32–55). New York: Continuum.

Martin, J. R. (2007). Construing knowledge: A functional linguistic perspective. In F. Christie & J. R. Martin (Eds.), *Language, knowledge and pedagogy: Functional linguistic and sociological perspectives* (pp. 34–64). New York: Continuum.

Morgan, C. (2005). Words, definitions and concepts in discourses of mathematics teaching and learning. *Language and Education, 19*(2), 103–117.

O'Halloran, K. L. (2005). *Mathematical discourse: Language, symbolism and visual images.* London: Continuum.

Rogers, M. P., Malancharuvil-Berkes, E., Mosley, M., Hui, D., & O'Garro Joseph, G. (2005). Critical discourse analysis in education: A review of the literature. *Review of Educational Research, 75*(3), 365–416.

Solomon, Y. (2009). *Mathematical literacy: Developing identities of inclusion.* New York: Routledge.

Tannen, D. (2007). *Talking voices: Repetition, dialogue, and imagery in conversational discourse.* New York: Cambridge University Press.

Veel, R. (1999). Language, knowledge, and authority in school mathematics. In F. Christie (Ed.), *Pedagogy and the shaping of consciousness* (pp. 185–216). London: Cassell.

Walshaw, M. (Ed.) (2004). *Mathematics education within the postmodern.* Greenwich, CT: Information Age.

Zevenbergen, R. (2003). Teachers' beliefs about teaching mathematics to students from socially disadvantaged backgrounds: Implications for social justice. In

L. Burton (Ed.), *Which way social justice and mathematics education?* (pp. 133–152). London: Praeger Publishers.

FURTHER READING

Fairclough, N. (2003). *Analyzing discourse: Textual analysis for social research.* New York: Routledge.

Jaworski, A. & Coupland, N. (Eds.) (1993). *The discourse reader.* New York: Routledge.

CHAPTER 8

PLAYING THE FIELD(S) OF MATHEMATICS EDUCATION

A Teacher Educator's Journey into Pedagogical and Paradoxical Possibilities

Kathleen Nolan

ABSTRACT

There is promising research on how inquiry-based classroom discourse in mathematics leads to enhanced student engagement and conceptual understanding (Walshaw & Anthony, 2008). While the notion of an inquiry-based classroom (what it looks and feels like) is interpreted in diverse ways across a variety of contexts, there are a few distinguishing features common to most. In general, inquiry-based pedagogy is an alternative view of teaching and learning, based primarily on theories of constructivism and characterized by classrooms where the focus is on constructing mathematical understanding through student investigation, collaboration, and communication (Cheeseman, 2008; Leikin & Rota, 2006). In university teacher education, however,

Unpacking Pedagogy: New Perspectives for Mathematics Classrooms, pages 153–173
153

mathematics educators struggle in their work with prospective teachers to help bring about a shift from traditional teacher-directed approaches to such inquiry-based classroom discourses (Makar, 2007; Manouchehri, 1998; Wilson, Cooney, & Stinson, 2005). This is due, in part, to the fact that the paradigm shift to inquiry-based classrooms also demands a tolerance for ambiguity, uncertainty, and negotiation—skills not generally acquired through years of traditional school mathematics experiences. Thus, mathematics teacher educators are confronted with numerous challenges and complexities as they work to inspire prospective teachers to embrace inquiry-based pedagogies, while also seeking to deconstruct what are perceived as firmly entrenched stereotypes and ideas about teaching (Weber & Mitchell, 1995).

My current scholarship as a mathematics teacher educator and researcher involves introducing prospective middle years teachers to alternative, inquiry-based pedagogy in mathematics. In recent research (Nolan, 2006), I proposed the existence of several classroom discourses that act to regulate the teaching and learning of mathematics, with the discourses often being in direct opposition to reform recommendations to effect change in mathematics classrooms. As Brown (2008) suggests, such regulative discourses can act as "cover stories" for why theories of alternative forms of pedagogy are not having a noticeable impact on classroom practice and new teacher development. At issue, of course, is the prospective teachers' own lack of experience as students with alternatives to traditional teacher-directed approaches in teaching and learning mathematics. Mathematics reform and teacher education initiatives must begin by acknowledging that the first step to changing the way prospective teachers *teach* is to change the way prospective teachers *learn* (Pereira, 2005). In other words, "unless teachers directly experience inquiry learning for themselves it is quite unlikely that they will be able to implement it in their classroom" (Carter & Richards, 1999, p. 70). In addition, I would argue that when prospective teachers have positive and self-affirming experiences with learning mathematics through alternative, inquiry-based pedagogy, they are less inclined to resist the traditional, familiar (and regulating) discourses of what it means to teach and to "cover" mathematics content.

The belief that prospective teachers require experience learning *through* (not merely *about*) inquiry-based pedagogy led me to design and teach a new course in a university undergraduate teacher education program. The course, entitled *Curricular Topics in Mathematics*, is a compulsory course taken by prospective middle years teachers in their second year of a four-year teacher education degree at a Canadian university. In the design of the course, I selected the content to focus on middle years and secondary school topics in geometry and statistics, while the pedagogy included a variety of inquiry-based approaches, such as mathematical investigations, problem-based learning, technology-integrated learning modules, and collaborative problem posing/ solving. The discussions in this chapter are based on data from a research project designed as a self-study narrative of my experience teaching this new course, using my own journal reflections (as the designer and instructor for the course), as well as the journal writings and assignments of the students enrolled in the course. In addition to journals, data included student auto-biographical "inventory" (survey) responses and final course evaluations. In this chapter, I present a reflexive narrative on my experience teaching this course, constructed through my own lens as a teacher (acknowledging my vulnerability in the face of student resistance and dissatisfaction) and through the lens of identifying with new teachers as they face similar student and classroom discourses of resistance. What makes this story unique is its candid approach to exploring why current mathematics teacher education programs generally have a superficial and temporary impact on reforming the teaching and learning of school mathematics. The story, told from my perspective as a mathematics teacher educator, highlights the need for a reconceptualized mathematics teacher education program that embraces the dynamic relationship between research, teaching, and learning.

Encouraging prospective mathematics teachers to make personal and professional transitions from traditional didactic teaching practices to inquiry-based approaches presents many challenges. While there has been valuable research to date on the nature of the transitions required for becoming teachers (Garcia, Sanchez, Escudero, & Llinares, 2006; Jaworski & Gellert, 2003; Klein, 2004; Ritchie & Wilson, 2000), the area is still relatively under-documented and under-explored in the research literature. In this chapter, I claim that the transitions of prospective teachers call for a drastic change of script in storylines for what it means to teach and learn mathematics. I propose that one way to understand and unpack the transitions is using Bourdieu's social field theory. In the following pages, I draw on aspects of Bourdieu's theory to explore the pedagogical and paradoxical possibilities of inquiry-based pedagogy in mathematics (teacher) education.

INTRODUCING IDEAS

General Introduction to Bourdieu's Social Field Theory

Bourdieu's social field theory consists of several key concepts and terms, including *practice, field, habitus, capital, doxa,* and *misrecognition.* In this section, I will define each of Bourdieu's key concepts and outline the network of relations between them.

The concepts of *field* and *habitus* are central to understanding social practice, since the two concepts are produced and reproduced in relation to each other through social practice. According to Bourdieu, the everyday decisions made in a social context (the field) shape, and are shaped by, a set of dispositions or tendencies (*habitus*) that are formed through the embodiment of an individual's (or a collective's) life history. In other words, "*[h]abitus* is a set of dispositions that are commonly held by members of a social group and these subjectively created attitudes, beliefs, and practices bind the members together so that they can identify and communicate with each other" (Mutch, 2006, p. 163). *Habitus* operates at various levels—in one's thoughts, actions, use of language, and even at the corporeal level, in terms of how one embodies experiences of structures and relations. A *field* refers to a social arena or context in which a network of these structures and relations is found (Grenfell, 2008). Bourdieu posits the existence of many possible fields, all "historically constituted areas of activity with their specific institutions and their own laws of functioning" (Bourdieu, 1990a, p. 87). As part of this network of relations, it is important to understand how the field and *habitus* are inseparable, mutually constituting, and complicit in each other in all social practices. Although *habitus* is often viewed as "merely a source of choices, rather than a lock step prescription" (Harker, 1990, p. 90), it is worth noting that a "well-formed" *habitus* (one that feels in synch with the social practices of the field within which one is immersed) will likely be quite durable and not so transposable when situated in other fields with different social practices.

A third key concept, and one that plays an important role in the relationship between field and *habitus*, is *capital*. Bourdieu describes two main forms of capital (economic and symbolic), but for the purposes of this text and its focus on mathematics classrooms, cultural capital (a form of symbolic capital) is most relevant. According to Grenfell (2008), cultural capital is basically a synonym for status (or position), and refers to the resources that one brings to (and/or has access to in) the field. When discussed in the context of education, cultural capital can include "commodities," such as one's level of education, classroom experiences, research knowledge, grades/marks, classroom management skills, comfort with the script or logic of the field (i.e., a good *habitus*–field match), and so forth. In short, cultural capital includes all the things that help people gain access to, and position themselves strategically within, fields.

Bourdieu conceives of *practice* as the relationship between the three key "thinking tools" (Bourdieu & Wacquant, 1989, p. 50) of field, *habitus*, and capital, expressing their interlocking relationship through the formula: [(*habitus*)(capital)] + field = practice (Bourdieu, 1990b; Grenfell, 2008). This formula illustrates three key aspects of social field theory: (1) social practice is shaped through the relationship between one's agency, one's position within a social context, and the rules and regulations of that social context; (2) both *habitus* and field are continually produced and reproduced in/through social practice and relative to each other (that is, the two are intimately connected through practice); and (3) *habitus* and capital have a direct bearing on each other—those with well-formed *habitus* are higher in cultural capital, and the opposite also holds true. Dimitriadis and Kamberelis (2006) express this dynamic relationship in stating:

A field is thus defined primarily in terms of the kinds of practices that are common within it and the kinds of capital that may accrue to individuals who engage in those practices, and secondarily as the kinds of social relations that develop as people work to acquire and maintain the kinds of capital with the most purchase in the field. (p. 67)

By understanding the dynamic roles of the three key concepts—field, *habitus,* and capital—and their complex interactions, social field theory can help illuminate issues of domination and reproduction in education. In fact, according to one interpretation of Bourdieu's theory and social reproduction, "the role of schools is to make students believe that the existing social relations are just and natural and in their interests" (Webb, Schirato, & Danaher, 2002, p. 113).

This brings me to two final social field theory concepts: *doxa* and *misrecognition. Doxa* is the set of core values and discourses of a social practice field that have come to be viewed as natural, normal, and inherently necessary, thus working to ensure that the arbitrary and contingent nature of these discourses is neither questioned nor even recognized. Such an uncritical acceptance of what constitutes normal, natural, and necessary is what Bourdieu refers to as "misrecognition" (Bourdieu, 1990a; Webb et al., 2002). According to Deer (2008), "*doxa* allows the socially arbitrary nature of power relations ... that have produced the *doxa* itself to continue to be misrecognized and as such to be reproduced in a self-reinforcing manner" (p. 121).

To gain insight into the relationships between Bourdieu's social practice constructs, consider his analogy of playing a game:

The concept of field allows the researcher to detail the context in which the action is taking place and to put boundaries around the place of action. The concept of capital explains who gets to play, why, and how. The concept of *habitus* can be used to describe and analyze the strategies of the players in each particular context. (Mutch, 2006, p. 170)

Bourdieu's view is that adjustment to the demands of a field requires a certain "feel for the game" (Bourdieu, 1990b, p. 66). As in a game (such as football), social fields are constructed with specific structures and rules, and the relative smoothness of the game/field often depends upon the players blindly accepting and following these rules, regardless of how arbitrary they might seem. As one continues to engage in the game, the rules seem natural and unquestionable to the players, resulting in a "feel for the game" that no longer requires the deliberate act of thinking carefully about each and every move before acting. Also, as with most games, social fields are competitive, with the players continually vying for better positions and more refined skills in the game.

Where this game analogy breaks down in relation to field theory, however, is that social fields are often not level playing fields; that is, those who begin the game with particular valued forms of capital and "well-formed" *habitus* are always at an advantage. In relation to schools and education, one can see how (and why) those who have a privileged position, and have learned how to play school well, have an investment in perpetuating and reproducing the logic and operations of the field, as is. A player's complicit and (re)productive role is discussed by Bourdieu (1990b):

> The earlier a player enters the game and the less he is aware of the associated learning... the greater is his ignorance of all that is tacitly granted through his investment in the field and his interest in its very existence and perpetuation and in everything that is played for in it, and his unawareness of the unthought presuppositions that the game produces and endlessly reproduces, thereby reproducing the conditions of its own perpetuation. (p. 67)

While Bourdieu's theory is primarily based in sociology, and frequently with reference to the social fields of gender, class, and the like, several researchers have adapted Bourdieu's theory to other fields, including educational policy (e.g., Mutch, 2006) and research practices (e.g., Grenfell, 2008). In addition, it is especially worth noting that several researchers have adapted aspects of Bourdieu's theory to explore issues in mathematics education (Noyes, 2004; Roth, Lawless, & Tobin, 2000; Wedege, 1999; Zevenbergen, 2000).

Following the next section of this chapter, I will further draw on aspects of Bourdieu's social field theory to reflect on specific examples of how my experience of teaching the new mathematics course presented both pedagogical and paradoxical possibilities for mathematics teacher education and mathematics education in general. In the discussion and interpretation section of this chapter, I grapple with these possibilities by describing two fields of play—the field of education in K–12 schools, particularly in mathematics classrooms (F1) and the field of university teacher education, particularly in mathematics curriculum courses (F2). In terms of these two fields, one could think of prospective teachers as having a social practice journey through F1 as a student, then F2 as a prospective teacher, and then back to F1 as a teacher. I propose that in each of these two fields, specific (but quite different) forms of *habitus* and cultural capital are valued and (re)produced. While one may expect and desire F1 and F2 to be compatible, even similar, fields of play (at least for the sake of smooth transitions for prospective teachers), my analysis and interpretation focuses primarily on the dissimilarities and misrecognitions that expose F1 and F2 as being, paradoxically, worlds apart.

APPLYING IDEAS

The "Real" and the "Reality" in Mathematics Teacher Education

As a teacher educator, I am often at a loss for words when I hear prospective teachers proclaim that the "real" learning experience within teacher education programs occurs during the practicum (or field) experiences. Prospective mathematics teachers claim that university courses do little to prepare them for the "realities" of school classrooms (Britzman, 2003; Furlong et al., 2000; Nolan, 2008). But if these "real" experiences and the "realities" of classrooms are actually misrecognitions of the values, rules, and *modus operandi* for what it means to teach, to learn, and to know mathematics, then what possibilities exist for reconceptualizing mathematics (teacher) education?

As prospective teachers seek to reconcile multiple and conflicting demands between teacher education programs and their practicum (field) experience in K–12 schools, they encounter competing discourses on the value of alternative, inquiry-based pedagogy in mathematics classrooms. While my voice adds to the reform discourse, encouraging them to embrace a full repertoire of pedagogical strategies, there are at least two other discourses resisting those reform calls, in terms of what is "real" and "reality" in K–12 mathematics classrooms. First, prospective teachers have already experienced many years of a traditional model of what it means to teach, learn, and know mathematics. Skovsmose (2008) aptly names this traditional model the "exercise paradigm" (p. 167)—a paradigm that is characterized by the teacher carefully executing example problem-solving exercises, followed by students performing several identical exercises in their notebooks. The exercise paradigm has been normalized in mathematics classrooms as the approach that is most effective at covering maximum curriculum content in a minimum period of time (Nolan, *forthcoming*). Even if prospective teachers were not successful and/or their experiences were not favorable within this paradigm, they frequently internalize the failure as their own lack of ability, or perhaps their teachers' inability to explain things clearly. Overall, repeated performance of the exercise paradigm has been the "real" experience of learning mathematics for many prospective teachers.

The second discourse at work in resisting reform is the practicum experience itself. When prospective teachers enter the practicum (field) experience, there is clear evidence to support that the exercise paradigm is alive and well in Canadian K–12 mathematics classrooms. Prior to their practicum experiences, I am generally optimistic that prospective teachers have a desire to "transcend the *habitual* in order to think the *possible* in mathematics classrooms" (Nolan, 2008, p. 164). Then, however, they enter a mathematics classroom characterized by discourses that closely reflect what they experienced as students. Research shows that field experience carries significant weight in the overall process of becoming a teacher because, for prospective teachers, it represents the "real" experience of teaching. As Britzman (2003) states: "Like most people in teacher education, [the prospective teachers] were deeply invested in the idea that experience is telling, that one learns by experience, by being there, and not by theories. If this were not the case, why have such a long internship?" (pp. 252–253).

Over the years, I have become acutely aware of these two discourses that work to regulate and normalize the traditional images of what it means to know and learn mathematics, and to be a mathematics teacher. The discourses are seldom disrupted in any significant manner in teacher education programs because, as Roth et al. (2000) suggest, "university teacher education classrooms continue to be filled with talk about strategies, techniques, and skills" (p. 2). In designing the new mathematics course, I acknowledged this predicament and made a deliberate effort to "walk the talk," hoping to disrupt the prospective teachers' versions of the "real" and the "reality" in the teaching and learning of mathematics.

Reflections on Teaching the New Mathematics Content Course

The new course, *Curricular Topics in Mathematics*, began quite smoothly. When I presented the students with the course outline, they were open and enthusiastic about the idea that mathematics teaching and learning could be so different from their own K-12 experience—that it could be characterized by problem posing/solving, meaningful student discussions, multiple approaches and ways of knowing, and reflective practices. In fact, even those prospective teachers who initially expressed much trepidation at the thought of "having to take another math course" quite readily embraced the key goal of this course: To work independently and collaboratively to experience, and develop an appreciation for, learning mathematics through a variety of inquiry-based approaches that highlight mathematical discourse and conceptual understanding.

After spending some time discussing the course outline, and then engaging the prospective teachers in a brief inquiry-based mathematical activity, I closed the first class by asking the students to complete a math inventory—an open-response survey to elicit initial reactions to the course content and goals. The following comments were drawn from these inventories:

Inventory Question: What part of the course outline interests you the most and why?
– New ways of looking at math
– Problem-based and investigative approaches interest me because I only really know one style of learning or studying math
– Learning ways to inspire learners without lectures or traditional methods of testing. . . . I am excited to observe methods that are out of the box
– Learning how to teach math in a more hands-on way
– Working with other classmates on problems to get a better understanding of the problem
– Digging deeper than just getting answers.

Inventory Question: Describe how/if you want your students' experiences of learning math to be different from your own experiences of learning math.
– More demonstration/hands-on learning rather than all from a text
– Make the topics fun & interesting . . . my math classes were always dry, repetitive, similar, and boring
– I would like to try and use different ways of explaining the same thing . . . math doesn't have to be hard
– Fun and creativity should be used . . . not just lectures, watch, example, and then on their own
– I would hope my students could fully understand concepts instead of memorization
– Not feeling it is just another thing to write in their notebook

In fact, one student wrote: "I once had a teacher who sat in a chair all class beside an overhead projector, and would only write notes and give a few examples. I never want to teach like that."

Initially, when I read these responses, I was optimistic that the prospective teachers were coming to this course with, at the very least, an open attitude toward challenging and changing their current perceptions on teaching and learning mathematics. I initiated our journey as a class by clarifying exactly what the course was about; that is, I explained that it was a mathematics content course *about* topics in geometry and statistics, taught *through* recommended inquiry-based pedagogical approaches. In other words, I would teach the mathematics content through the pedagogical approaches advocated in the reform literature and new curriculum documents—the same approaches prospective teachers would be expected to use to teach the very same content in their own (future) classrooms. As discussed previously, rather than talking, reading, and writing *about* these non-traditional approaches, I endeavored to "walk the talk" by teaching the mathematics content *through* them.

As the course progressed, however, several incidents occurred that began to make my experience as the instructor quite disheartening. For example, early in the course, students were asked to complete a self-directed web-based module as a review of introductory statistical terms and analyses. There was no official assignment to hand in for grading upon completion of this module. Instead, the intent was for students to work through the material in the module, to self-assess as to which content areas required some attention, and then, on their own initiative, to delve into the designated websites to learn about these areas. I quickly surmised that students did not complete this self-directed module when they were unable to understand and complete the course work that followed, and built upon, the content of the module. My reading of this situation was that student learning was seriously impaired by their lack of engagement with course work that was required to *know* but was not *graded.* My discomfort with the situation reached an impasse when students reflected in their journals (and later on final course evaluations) that they were asked to do work that they "were not taught."

Approximately halfway through the course, a similarly disheartening and frustrating experience occurred. Students wrote in their assignment journals that they wished they could learn more about *how to teach math,* instead of spending so much time actually doing the math. In response to this desire expressed through their journals, I wrote the following in my own personal journal, where I pondered the question: If I "walk the talk," how will I ensure they "get it"?

> They just don't get it. At the most basic description of this course, it is concerned with learning math *content* because evidence suggests that middle years prospective teachers don't understand the content of the mathematics curriculum they will soon teach. They need the content taught to them again. Why? Well, all the research I've read and done myself suggests that it's because the ways in which they experienced (learned?) the content as K–12 school students did not connect with them, and so they generally did not reach an understanding of the concepts—only the procedures, and only for a short time ("until the test was over," they would report).
>
> Now, in this class, I've made it abundantly clear that the course is about learning that same middle years and secondary math content, but this time through more constructivist, collaborative, investigative, and problem solving/posing ways. Of course, I've not only *told* them this, but they are in the middle of experiencing it. And yet, they're telling me that they want to learn more about *how to teach math . . .* what else can I do? Do they require *notes* on how to teach in order for them to make any association between what they are actually engaging in now and what they will be doing in their own future classrooms

as teachers? What must I do to reach them and make a tear in their fabric of "what it means to teach and learn" —the fabric that is so neatly folded and sitting in the corner, ready for practicum?

This just doesn't seem worth it. Why did I think that somehow they would each have an "aha" moment, exclaiming "Gee, if only I had learned math this way before, I would have understood it!"?

I admit now that my illusions of grandeur involved the notion that I would teach a few classes through inquiry-based approaches, expecting and accepting some initial student discomfort and angst, but then soon enough a few students would experience "aha" moments and suddenly understand what I was trying to do. They would then proceed to share their insights with others and the course would become a smooth experience for all. I clung to the hope that these "aha" moments (when they finally arrived and spread like wild fire through the classroom) would be a "cure" for the resistance I was encountering. I reasoned that my stance was worth the effort and that, when it was time for final course evaluations, I could count on a confession rather than an attack. Not so. As it turned out, the open-ended nature of the mathematics activities introduced ambiguity and uncertainty, and this was found to be incredibly frustrating for the prospective teachers. The collaborative nature of problem solving that was emphasized in the course activities injected a dose of skepticism into their previously constructed world of competitive mathematics, and this was foreign and disconcerting for prospective teachers. Experiencing a view that knowledge is socially constructed (and not merely residing in the heads of all teachers everywhere) only served to fuel the fire in their accounts of my incompetence and inability to provide "straightforward instruction."

Overall, my good days as an instructor became fewer and further between. The students were not willing to work on their own or in small groups without being assured that their reward would come in the form of teacher-supplied right answers prior to leaving class each day. If they were asked to work on a computer-integrated learning module, they would assume that the "official" part of class was over and they would chime in unison, "If that's all there is, then can we leave now and finish this later?" In spite of many hours of deliberate preparation on my part to provide appropriate (but not overly prescriptive) direction and scaffolding for their learning, the students continued to express dissatisfaction with most aspects of the course. At the end of the course, the following general comments and suggestions for improving the course were written on the evaluation forms:

- More teaching, less us figuring it out for ourselves
- More explanation, more instruction, more examples
- More straight forward instruction

- Material did not seem relevant to middle years math and if it was, it wasn't explained well enough
- This class took up way too much time . . . and the marking never reflected the effort I put in
- Too much time into pointless activities
- I didn't really learn stuff that would help me as a becoming teacher
- Teach me more things that I can use in the classroom
- I was expecting this class to present methods of teaching math

And, as one student summarized, "I feel I have learned absolutely nothing and that this class was a complete waste of time, as I taught myself everything, since I was just given a sheet and thrown in front of a computer to teach myself."

At the time of designing and teaching the course, I recognized the challenges before me. I was asking these prospective teachers to make the transition from experiencing total dependence on the teacher and textbook as the authorities, including quite prescriptive K–12 classroom structures for learning—their familiar *habitus*—to in(ter)dependence and self-motivation in learning. I was not unsympathetic to the resistance I experienced from the prospective teachers, nor to the paradoxical nature of their desires: They *wanted* (in theory) to be open to embracing a reconceptualized mathematics classroom (as advocated in F2, my mathematics course), but only if the ideas came in a neat and tidy package that could merely replace (in practice) their tried-and-true package (as shaped in F1, their K–12 mathematics classrooms).

Intersections Between Research, Teaching, and Learning

Naturally, there were also a few positive outcomes of teaching this course, and the paradoxical nature of my experience was certainly not lost on me. Teaching this course awakened me to the parallels existing between the risks that I hope prospective teachers will take by teaching through inquiry-based pedagogy in their future middle years mathematics classrooms and the risks that I was taking in this course by embracing the complexity and ambiguity of these same alternative methods in mathematics teacher education. The paradox is, however, that while teacher educators are expected to prepare K–12 teachers to embrace the ambiguity, uncertainty, and unpredictability associated with the messy realities of teaching, they feel pressure themselves to reduce (and even eliminate) these messy realities in their *own* teaching practices. Unfortunately, the field of teacher education appears entrenched in a modernist paradigm, as described by Smits and Friesen (2002):

> The teaching about and learning of teaching has been conceived and practiced as the correct application of universalized and abstracted norms and rules.... But when we attempt to shed this model of teacher education and what it implies for our teaching work, we are somewhat at a loss to define clearly—either in theoretical or practical terms—what it is we both represent and profess. This has implications for how our students perceive and receive our work in the university classroom, and how and what we represent as legitimate teacher education knowledge. (p. 80)

In other words, the subtext of teacher education discourse reads that teacher educators are expected to "have it all figured out," thus creating the façade that teaching and learning are actually *not* messy or ambiguous processes. Such a façade does not create spaces for embracing the learning of teacher educators as a necessary ingredient for the growth and reconceptualization of teacher education programs. It appears that it is perfectly acceptable for teacher educators to be learners in their *research*, but not in their *teaching*. Thus, current teacher education programs are viewed as stagnant places of "training" and "preparation," rather than dynamic spaces in which to explore the relationships between research, teaching, and learning (Brew, 2006).

To the contrary, I actually think of my experience with this course as a critical moment of intersection between my research and my teaching. I believe that a course such as this one should provide a "facilitating context" (Lattuca, 2002, p. 736) for my research—that is, the course should serve as an arena for exploring how my research, teaching, and learning all intersect and inform/shape one another. In Nolan (2005), I call for a dialectic relationship between the acts of researching *and* teaching *and* learning, recognizing that if we value constructivist and deconstructivist approaches to these academic "acts," then the lines drawn between them—the lines drawn in the (s)and—are fabricated and unstable. Cochran-Smith and Lytle (2006) echo this call, stating that "there are critical relationships between teacher learning and student learning; that when teachers learn differently, students learn differently" (p. 692).

An environment where I am encouraged to learn, and to learn with my students, is one in which I would feel safe to reconcile my own conflicting demands and competing discourses on the possibilities and promises of alternative pedagogy in mathematics.

Viewing this course through the lens of Bourdieu's theory holds considerable promise for understanding and processing the experience. In the language of Bourdieu's theory, it could be said that my own dispositions and tendencies toward teaching and learning mathematics (my *habitus*) have been strategically shaped in and through my own school experiences (F1), as a student and then as a teacher. Now, however, as a teacher educator, I am motivated to reshape my *habitus* so that it matches a different set of "rules" and practices in F2. I seek to reconceptualize the field of university mathematics teacher education as a social arena where both teacher educators and prospective teachers strive to recognize that which is continually misrecognized in mathematics classrooms. In this reconceptualized field of mathematics teacher education, characterized by theories of postmodernism and inquiry-based learning, I am experiencing discomfort in my own transition process of transposing and reshaping my *habitus*. Unfortunately, the prospective teachers' responses to my struggles and discomfort with my own transitions eventually elicited a desire within me to give in to their resistance and provide them with what they wanted—unambiguous, straightforward tips and techniques for how to teach mathematics. By doing so, however, I would be further enabling a *doxic* attitude and a propensity for misrecognizing the reproductive role of schools and teacher education.

DISCUSSION AND INTERPRETATION

In this discussion and interpretation section, I seek to further connect aspects of Bourdieu's theory to understand the reflections already put forth in this text. As presented earlier, I propose two possible fields of play: the field of education in K–12 schools, particularly in mathematics classrooms (F1), and the field of university teacher education, particularly in mathematics curriculum courses (F2). In these two fields, the unwritten rules of the game interact with, and shape, the *habitus* and cultural capital that each social agent brings to, and forms in, the field.

It should be noted that, in general, fields of social practice are dynamic, multiple, and overlapping. In the specific case of this text, F1 and F2 are just two of the many fields that make up the larger society or social space in which prospective teachers find themselves at various points in their lives. To speak of prospective teachers in/of these two fields (moving through F1 as a student then F2 as a prospective teacher, and then back to F1 as a teacher) as if no other fields impact or influence their journeys would be entirely misleading and incorrect. For the purposes of this text and analysis, however, I have limited my discussion to these two fields because my interest lies in understanding the nature of the interactions and transitions between the practices of mathematics teacher education programs and K–12 mathematics classrooms.

As introduced previously, the analogy of playing a game—and thus seeing F1 and F2 as fields of play—helps illustrate the dynamic relationship between field, *habitus*, and capital in any social space. A person's understanding of, and feel for, the rules of the game—while seldom explicit and overt—constitute the *habitus*. With regard to the particular context of prospective middle years mathematics teachers, the practices in K–12 schools (F1) over many years as a student have shaped their *habitus* and cultural capital. This means that when prospective teachers return to F1 (after a short period of time in F2), the feel for the game is remarkably familiar and comfortable. For prospective teachers, F2 is a blip on the radar as they pass from F1 (as a student) through F2 (as a prospective teacher) and back to F1 (as a teacher). The dispositions formed and shaped in F1 as a student are durable and, without significant and long-lasting intervention of different *habitus* and/or different field "rules," F1 remains a good fit.

There is at least one significant difference with the *habitus*–field fit in F1 the second time around, however. This time prospective teachers bring more cultural capital to F1, thereby improving their positioning, or status, in F1. Now they have a university degree, a form of cultural capital that positions them in the powerful role of teacher. Cultural capital is a form of power and, as with most forms of power, it works to define a person's trajectory, based on how well it "works" in the field. Grenfell (2008) proposes that those with higher capital are also those who have more power in setting the rules of the game, implying that the *habitus* of the dominant group is the one most "enduring" in any social practice.

Since cultural capital explains who gets to play the game, as well as why and how the game gets played, it is not difficult to see then why the *habitus* and cultural capital a prospective teacher brings to F1 the second time around (as a teacher) results in an even more comfortable fit than it did when he or she was a student. They have developed a feel for the game in F1, which "takes prolonged immersion to develop" (Maton, 2008, p. 54). On the other hand, F2—the field of mathematics teacher education and their experiences of my curriculum courses—does not feel so comfortable. My analysis suggests that the *doxa* of their "feel for the game" in F1 comes under the spotlight in F2, where the misrecognitions become my target for deconstruction.

As an illustrative example of F1 *doxa*, consider prospective teachers' images of what it means to be a good mathematics teacher and the stories that have shaped these images. For example, the stories of the good mathematics teacher go something like this: "I will explain things clearly to my students"; "I must be able to effectively manage my classroom"; "I should know and be ready to provide the answers to the assigned mathematics tasks as a sign of competence." Unfortunately, many of the images of the good teacher focus on technical, rational concerns that pervade the fields of teaching and teacher education (Burton, 2004; Moore, 2004; Roth et al., 2000). In my mathematics course, I attempted to disrupt these images of the good teacher and, as a result, my stories conflicted on a level that served to widen the gap between prospective teachers' images of good teachers and what I was promoting (and attempting to model) as good teaching. Brown and McNamara (2005) suggest that prospective teachers need to "re-script their personal storyline to accommodate external demands that disrupt their original aspirations of what it is to be a teacher" (p. 28). Parallels can be drawn between these images of the good teacher (of mathematics) and Bourdieu's description of the "good player" (of the game): "The *habitus* as the feel for the game is the social game embodied and turned into a second nature. Nothing is simultaneously freer and more constrained than the action of the good player" (Bourdieu, 1990a, p. 63). Like the description of the good player, the image of the good mathematics teacher may be seemingly freeing, yet unconsciously constrained through its tight script in the field.

On the surface, the goals of my course in F2 are enticing and desirable, but prospective teachers' dispositions (shaped through their K–12 student experiences in F1) are not a good fit for the course. In accepting the *doxa* of F1, prospective teachers see no real benefit in forming new dispositions in F2 that will not be well matched for the rules of the game back in F1, when they become teachers. Even though *habitus* and field are dynamic—always evolving, always partial and never a *perfect* match for each other—a person will be most comfortable in a field when his or her *habitus* is a good fit for the logic and operation of the field. Even though F2 may be uncomfortable for prospective teachers, fortunately (for them), it is a brief detour between F1 as a student and F1 as a teacher. Instead of recognizing the possibilities in/for reshaping their *habitus,* and, in turn, using that new *habitus* to reshape the field of F1, they misrecognize the mutually constitutive nature of *habitus* and field.

It is worth clarifying that my use of Bourdieu's language of "well-formed" and "good *habitus*–field fit" is not intended to imply that prospective teachers are comfortable with the subject of mathematics and/or with teaching it. On the contrary, their well-formed ways of acting, thinking, and feeling may seem *natural* to them without meaning that they have been voluntarily structured, or even necessarily based on favorable experiences. Having a feel for the game in F1 does not translate into an assumption of enjoyment, success, or the development of positive dispositions toward the game. Having a feel for the game does imply, however, that the prospective teachers *did* develop dispositions, and, in their *doxic* state, these dispositions are durable and resistant to change.

CONCLUDING THOUGHTS ON PEDAGOGICAL AND/OR PARADOXICAL POSSIBILITIES

Recall that on the initial course inventory, one student made the following comment: "I once had a teacher who sat in a chair all class beside an overhead projector, and would only write notes and give a few examples. I never want to teach like that." It is not unusual for prospective teachers, especially those who had unsuccessful and/or unfavorable experiences, to rebuke the teachers who taught them mathematics, promising to never teach that way when they are teachers. They vow to make valiant efforts to generate meaningful mathematics experiences through group work and inquiry-based tasks that value student ideas and voice. As it turns out, however, this is not a trivial task when they become faced with the task of reconciling conflicting demands amid the pull of normative discourses. Such normative discourses often succeed in marginalizing and misrecognizing the promises of alternative discourses. The "exercise paradigm" (Skovsmose, 2008) is an example of one such normative discourse that has come to attain *doxic* status—teachers misrecognize the potential for peeling away the layers on the image of the good teacher as one who demonstrates competence and expert knowledge through "straightforward instruction." Bourdieu's social field theory helps view these competing and conflicting demands on prospective teachers in a new light, with an understanding that the passive act of *wanting* to change one's *habitus* is easier said than done when the rules of the playing field continue to appear unaltered in any significant manner. The set of dispositions and tendencies constituting one's *habitus* are durable, and, since they have been formed through prolonged exposure in the field of K–12 schools (F1), merely stating "I never want to teach like that" only underscores the power and paradox of misrecognitions.

My use of Bourdieu's social field theory in the context of playing the field(s) of mathematics education is not intended to be deterministic and overly structural, even though the snapshot in time that I have provided appears as such. Through his key thinking tools, Bourdieu portrays the relationship between field, *habitus*, capital, and social practice as necessarily a reflexive one, since there is no place outside these systems where one can obtain a neutral, detached perspective. In fact, "Bourdieu asks researchers to adopt a reflexive attitude towards our practices, reflecting upon how forces such as social and cultural background, our position within particular fields and intellectual bias shape the way we view the world" (Webb et al., 2002, p. xiv). This is where possibilities and hopes reside—in the reflexive approach to understanding and acknowledging that what is needed to reveal "the hidden workings of *habitus* [is] a political form of therapy enabling social agents to understand more fully their place in the social world" (Maton, 2008, p. 59).

Thomson (2008) shares a sign of hope, in terms of *habitus*–field reshaping, in writing that "social agents can experience change in fields when there is a disjunction between their *habitus* and the current conditions within the field" (p. 79). In mathematics teacher education, there is a call to create the conditions for prospective teachers to experience such disjunctions in their current comfortable *habitus*–field fit in K–12 mathematics education. In addition to initiating a reshaping of their own *habitus*, the disjunctions will, in turn, prompt a reshaping of the field itself, since *habitus* and field are mutually constitutive and complicit in each other. The challenge, however, lies in encouraging prospective teachers (and teacher educators) to take risks and consider trying an uncomfortable *habitus* on for size. It is likely that such risks will eventually transpire when teachers, prospective teachers, and teacher educators all recognize that "the game that is played in fields has no ultimate winner, it is an unending game, and this always implies the potential for change at any time" (Thomson, 2008, p. 79). Mathematics teaching and teacher education are in need of reconceptualization (Bullock, 2007; Lerman, 2001; Simon, 1997), and that is what makes the unending game of playing the field(s) of mathematics education worthwhile.

In this chapter, I have taken a reflexive approach toward understanding my *habitus* as a teacher educator in the field of mathematics teacher education. Through Bourdieu's theory and a self-study approach to my research and teaching, I am recognizing the paradoxical nature of my role as a teacher educator—one that gives me cause to reflect on my own complicity in reproducing a desire to reduce (the discomfort associated with) the ambiguities and complexities of teaching and teacher education. In doing so, however, I have also highlighted possibilities for a reflexive and hopeful approach to recognizing that which has been systematically misrecognized in mathematics (teacher) education for so long. The promising aspect of using Bourdieu's theory in this context lies in how the three key elements of *habitus*, capital, and field have the potential to become dynamic dispositions, positions, and spaces. With permeable and shifting boundaries in both playing fields of mathematics education, the pedagogical possibilities may begin to emerge through a steadily increasing area of reflexive overlap between F1 and F2. In other words, when new pedagogies of mathematics education and of mathematics teacher education are viewed as inextricably intertwined (Russell & Loughran, 2007), then teachers and teacher educators may embark on a new journey of understanding our pedagogical and paradoxical actions. After all, "the study of human lives would not be worth the trouble if it did not help agents to grasp the meaning of their actions" (Postone, LiPuma, & Calhoun, 1993, p. 6).

NOTE

1. The term *middle years* refers to students aged 11–13; in Canada, this coincides with grades 6–8 (upper primary school) within our Kindergarten (K)–12 school system.

REFERENCES

Bourdieu, P. (1990a). *In other words: Essays toward a reflexive sociology* (Trans: M. Adamson). Cambridge: Polity Press.

Bourdieu, P. (1990b). *The logic of practice* (Trans: R. Nice). Cambridge: Polity Press.

Bourdieu, P., & Wacquant, L. (1989). Towards a reflexive sociology: A workshop with Pierre Bourdieu. *Sociological Theory, 7*(1), 26–63.

Brew, A. (2006). *Research and teaching: Beyond the divide.* UK: Palgrave MacMillan

Britzman, D. (2003). *Practice makes practice: A critical study of learning to teach (revised edition).* New York: State University of New York Press.

Brown, T. (2008). Comforting narratives of compliance: Psychoanalytic perspectives on new teacher responses to mathematics policy reform. In E. de Freitas & K. Nolan (Eds.), *Opening the research text: Critical insights and in(ter)ventions into mathematics education* (pp. 97–109). New York: Springer.

Brown, T., & McNamara, O. (2005). *New teacher identity and regulative government: The discursive formation of primary mathematics teacher education.* New York: Springer.

Bullock, S. (2007). Finding my way from teacher to teacher educator: Valuing innovative pedagogy and inquiry into practice. In T. Russell & J. Loughran (Eds.), *Enacting a pedagogy of teacher education: Values, relationships and practices* (pp. 77–94). New York: Routledge.

Burton, L. (2004). *Mathematicians as enquirers: Learning about learning mathematics.* Dordrecht: Kluwer Academic Publishers.

Carter, R., & Richards, J. (1999). Dilemmas of constructivist mathematics teaching: Instances from classroom practice. In B. Jaworski, T. Wood, & S. Dawson (Eds.), *Mathematics teacher education: Critical international perspectives* (pp. 69–77). UK: Falmer Press.

Cheeseman, L. (2008). Does student success motivate teachers to sustain reform-oriented pedagogy? In M. Goos, R. Brown, & K. Makar (Eds.), Proceedings of the 31st Annual Conference of the Mathematics Education Research Group of Australasia (Vol. 1, pp. 125–130). Brisbane, AU: MERGA.

Cochran-Smith, M., & Lytle, S. (2006). Troubling images of teaching in No Child Left Behind. *Harvard Educational Review, 76*(4), 668–697.

Deer, C. (2008). Doxa. In M. Grenfell (Ed.), *Pierre Bourdieu: Key concepts* (pp. 119–130). Stocksfield, UK: Acumen Publishing.

Dimitriadis, G., & Kamberelis, G. (2006). *Theory for education.* New York: Routledge.

Furlong, J., Barton, L., Miles, S., Whiting, C., & Whitty, G. (2000). *Teacher education in transition: Re-forming professionalism.* Buckingham: Open University Press.

Garcia, M., Sanchez, V., Escudero, I., & Llinares, S. (2006). The dialectic relationship between research and practice in mathematics teacher education. *Journal of Mathematics Teacher Education, 9,* 109–128.

Grenfell, M. (Ed.) (2008). *Pierre Bourdieu: Key concepts.* Stocksfield, UK: Acumen Publishing.

Harker, R. (1990). Bourdieu—education and reproduction. In R. Harker, C. Mahar, & C. Wilkes (Eds.), *An introduction to the work of Pierre Bourdieu* (pp. 86–108). London: MacMillan.

Jaworski, B., & Gellert, U. (2003). Educating new mathematics teachers: Integrating theory and practice, and the roles of practising teachers. In A. Bishop, M.A.

Clements, C. Keitel, J. Kilpatrick, & F. K. S. Leung (Eds.), *Second International Handbook of Mathematics Education (Part Two)* (pp. 829–875). Dordrecht: Kluwer Academic Publishers.

Klein, M. (2004). The premise and promise of inquiry based mathematics in pre-service teacher education: A poststructuralist analysis. *Asia-Pacific Journal of Teacher Education, 32*(1), 35–47.

Lattuca, L. (2002). Learning interdisciplinarity: Sociocultural perspectives on academic work. *The Journal of Higher Education, 73*(6), 711–739.

Leikin, R., & Rota, S. (2006). Learning through teaching: A case study on the development of a mathematics teacher's proficiency in managing an inquiry-based classroom. *Mathematics Education Research Journal, 18*(3), 44–68.

Lerman, S. (2001). A review of research perspectives on mathematics teacher education. In F.-L. Lin & T. J. Cooney (Eds.), *Making sense of mathematics teacher education* (pp. 33–52). Dordrecht: Kluwer Academic Publishers.

Makar, K. (2007). Connection levers: Supports for building teachers' confidence and commitment to teach mathematics and statistics through inquiry. *Mathematics Teacher Education and Development, 8*, 48–73.

Manouchehri, A. (1998). Mathematics curriculum reform and teachers: What are the dilemmas? *Journal of Teacher Education, 49*(4), 276–286.

Maton, K. (2008). Habitus. In M. Grenfell (Ed.), *Pierre Bourdieu: Key concepts* (pp. 49–65). Stocksfield, UK: Acumen Publishing.

Moore, A. (2004). *The good teacher: Dominant discourses in teaching and teacher education.* London: RoutledgeFalmer.

Mutch, C. (2006). Adapting Bourdieu's field theory to explain decision-making processes in educational policy. In V. Anfara & N. Mertz (Eds.), *Theoretical frameworks in educational research* (pp. 155–174). Thousand Oaks, CA: Sage.

Nolan, K. (2005). Publish or cherish? Performing a dissertation in/between research spaces. In R. Barnett (Ed.), *Reshaping Universities: New relationships between research, scholarship and teaching* (pp. 119–135). United Kingdom: Open University Press.

Nolan, K. (2006). *A socio-cultural approach to understanding pre-service teachers' negotiated journeys through theory/practice transitions.* Paper presented at the 2006 Annual Meeting of the American Educational Research Association (AERA), San Francisco, CA, 7–11 April.

Nolan, K. (2008). Imagine there's no haven: Exploring the desires and dilemmas of a mathematics education researcher. In T. Brown (Ed.), *The psychology of mathematics education: A psychoanalytic displacement* (pp. 159–181). Rotterdam: Sense Publishers.

Nolan, K. (*forthcoming*). "For the sake of time" and other stories to teach by: Dis/positioning regulative discourses in secondary mathematics teacher education.

Noyes, A. (2004). (Re)producing mathematics educators: A sociological perspective. *Teaching Education, 15*(3), 243–256.

Pereira, P. (2005). Becoming a teacher of mathematics. *Studying Teacher Education, 1*(1), 69–83.

Postone, M., Lipuma, E., & Calhoun, C. (1993). Introduction: Bourdieu and social theory. In C. Calhoun, E. LiPuma, & M. Postone (Eds.), *Bourdieu: Critical perspectives* (pp. 1–13). Chicago: The University of Chicago Press.

Ritchie, J., & Wilson, D. (2000). *Teacher narrative as critical inquiry: Rewriting the script.* New York: Teachers College Press.

Roth, W-M, Lawless, D., & Tobin, K. (2000). Towards a praxeology of teaching. *Canadian Journal of Education, 25*(1), 1–15.

Russell, T., & Loughran, J. (Eds.). (2007). *Enacting a pedagogy of teacher education: Values, relationships and practices.* New York: Routledge.

Simon, M. (1997). Developing new models of mathematics teaching: An imperative for research on mathematics teacher development. In E. Fennema & B. Nelson (Eds.), *Mathematics teachers in transition* (pp. 55–86). Mahwah, NJ: Lawrence Erlbaum Associates.

Skovsmose, O. (2008). Mathematics education in a knowledge market: Developing functional and critical competencies. In E. de Freitas & K. Nolan (Eds.), *Opening the research text: Critical insights and in(ter)ventions into mathematics education* (pp. 159–174). New York: Springer.

Smits, H., & Friesen, D. (2002). From "common grounds" to the "rough ground" of teacher education: Experiencing teacher education as a collaborative practice. In H. Christiansen & S. Ramadevi (Eds.), *Reeducating the educator: Global perspectives on community building* (pp. 71–89). New York: SUNY.

Thomson, P. (2008). Field. In M. Grenfell (Ed.), *Pierre Bourdieu: Key concepts* (pp. 67–81). Stocksfield, UK: Acumen Publishing.

Walshaw, M., & Anthony, G. (2008). The teacher's role in classroom discourse: A review of recent research into mathematics classrooms. *Review of Educational Research, 78*(3), 516–551.

Webb, J., Schirato, T., & Danaher, G. (2002). *Understanding Bourdieu.* London: Sage Publications.

Weber, S., & Mitchell, C. (1995). *That's funny, you don't look like a teacher! Interrogating images and identity in popular culture.* London: The Falmer Press.

Wedege, T. (1999). To know or not to know—Mathematics, that is a question of context. *Educational Studies in Mathematics, 39*(1/3), 205–227.

Wilson, P., Cooney, T., &. Stinson, D. (2005). What constitutes good mathematics teaching and how it develops: Nine high school teachers' perspectives. *Journal of Mathematics Teacher Education, 8,* 83–111.

Zevenbergen, R. (2000). "Cracking the code" of mathematics classrooms: School success as a function of linguistic, social, and cultural background. In J. Boaler (Ed.), *Multiple perspectives on mathematics teaching and learning* (pp. 201–223). Westport, CT: Ablex Publishing.

FURTHER READING

Grenfell, M. (Ed.) (2008). *Pierre Bourdieu: Key concepts.* Stocksfield, UK: Acumen Publishing.

Noyes, A. (2004). (Re)producing mathematics educators: A sociological perspective. *Teaching Education, 15*(3), 243–256.

Webb, J., Schirato, T., & Danaher, G. (2002). *Understanding Bourdieu.* London: Sage Publications.

SECTION 3

INTEGRATED APPROACHES TO PEDAGOGY

CHAPTER 9

LIFE IN MATHEMATICS

Evolutionary Perspectives
on Subject Matter

Moshe Renert and Brent Davis

ABSTRACT

Edgardo Cheb-Terrab is an applied mathematician who specializes in developing algebraic algorithms for Maple, a mathematical software package for symbolic computation. His algorithms solve classes of problems, including differential equations and special functions. In 1999, Cheb-Terrab used his software to investigate solutions of Abel equations—a class of first-order non-linear differential equations that was first described by the Norwegian mathematician Niels Abel in the 1820s. By the end of the 20th century, the mathematics research literature contained solutions to nearly 40 types of Abel equations; each of these types was thought to require its own method of solution. Cheb-Terrab showed that all these types are special cases of an 8-parameter hyper-class. By devising a computer algorithm for solving all equations of this hyper-class, he expanded the range of solvable Abel equations far beyond what was thought possible. These results could not have been derived without a computer.

Even though one may expect such innovation to be greeted enthusiastically, Cheb-Terrab's work has met with suspicion and even dismissal by mainstream

Unpacking Pedagogy: New Perspectives for Mathematics Classrooms, pages 177–199
Copyright © 2010 by Information Age Publishing
All rights of reproduction in any form reserved.

algebraists. "For many of them, it was heresy; surely a computer cannot solve problems that were impenetrable to such great mathematicians as Abel and Liouville" (E.S. Cheb-Terrab, personal communication, November 2, 2008). Ironically, Cheb-Terrab's censure by the research community came at a time when thousands of mathematicians, engineers, and physicists were already using his algorithms, and verifying the solutions obtained, in a variety of applications. It took four years, and a lengthy review process, to publish the findings (Cheb-Terrab & Roche, 2003).

Cheb-Terrab has since developed innovative computer-based solutions to other long-standing problems in algebra. Yet he regularly confronts obstacles to acceptance. As Cheb-Terrab (personal communication, December 13, 2008) noted, "[m]athematicians have received with discomfort almost every algorithm that I developed which seemed to challenge 'established truths'; but, in fact, these established truths were never anything more than incomplete truths holding back progress in their fields."

The issue of what constitutes acceptable mathematics innovation points to a prevalent orthodoxy among mathematicians around the question "What is mathematics?" Even though today's digital technologies enable new mathematical understandings, many mathematicians are unwilling to accept computerized solutions as "real mathematics." For them, the only true mathematics is that which manifests in the time-honored mechanisms of formal proof. As Cheb-Terrab's example illustrates, these mathematicians, in their strict conformity to traditional modes of mathematical knowledge production, may be stunting the evolution of mathematics and contributing to stagnation of mathematical research.

Likewise, we believe that part of the blame for the current stagnation of mathematics education and pedagogy can be attributed to a shared orthodoxy among educators around the question "What is mathematics?" From an evolutionary perspective, we understand the term *orthodoxy* as referring to rigid adherence to a particular worldview, and refusal to acknowledge and participate in the evolution of consciousness. Mathematics pedagogues often perceive mathematics as a treasured, monolithic, even sacred body of knowledge, which must be preserved and passed on to future generations. This perception of mathematics often leads to a model of instruction that centers on transmission of stable knowledge.

This chapter explores the range of worldviews that respond to the question "What is mathematics?" We will use the evolutionary frameworks of complexity science and integral philosophy to analyze the stages through which conceptions of mathematics have evolved to date, and where they are likely to evolve next. We examine barriers to the evolution of teachers' views on mathematics, and some approaches from the authors' experience to overcome these barriers. We conclude with a discussion of the implications that an evolutionary view of mathematics holds for pedagogy, and, in particular, the need for educators to balance stable and emergent dimensions of mathematics in their instruction.

INTRODUCING IDEAS

Evolutionary Theories: Complexity Science and Integral Philosophy

Complexity science is the study of self-maintaining, evolving systems. These systems arise from the co-dependent interactions of autonomous agents, and evolve through the non-linear dynamic processes of self-organization (autopoiesis) and emergence. Complexity science has been used to study such varied physical phenomena as anthills, brains, and eco-systems. Integral philosophy seeks to extend the findings of complexity science regarding evolution in the exterior realm of nature, and to apply them to evolution in the internal realms of self and culture. Indeed, one of integral philosophy's main claims is that essentially the same patterns govern evolution in the three realms of nature, self, and culture (e.g., McIntosh, 2007; Wilber, 1995, 2006). Integral philosophy supports this claim by constructing an expansive framework of reality, an integral map that encompasses all three realms. To achieve this construction, integral philosophy draws on results and insights from complexity science, developmental psychology, and Western and Eastern philosophical traditions.

The roots of integral philosophy can be traced to Hegel, who posited that every domain of reality develops in a dialectical process, wherein synthesis transcends and partially preserves the conflicting dualisms inherent in any given thesis and antithesis. Hegel saw reality as an ongoing process of "becoming," and understood knowledge and consciousness as developmental processes. Dialectic thinking has since been applied in many domains. Marx, for example, interpreted social evolution as a dialectic between techno-economic means and class structure.

Perhaps most pertinent to educators are the insights of developmental psychologists, who were the first to describe how consciousness evolves dialectically through distinct and universal stages of development. In 1911, James Mark Baldwin outlined the first stage model for the dialectical development of human consciousness. It consisted of five distinct stages: *pre-logical, quasi-logical, logical, extra-logical,* and *hyper-logical.* Baldwin's work had a pivotal influence on Piaget's subsequent theories of cognitive development in children. Piaget's work, in turn, inspired numerous psychologists who offered stage models for different aspects of consciousness development (e.g., Kohlberg, Loevinger, Maslow). In his study of the psychological development of the self, Kegan (1994) observed that developmental stages exhibit an alternating pattern of subject-object mutuality, in which the subject at each stage of development becomes the object of another subject in the stage that follows.

Graves (1970) and his followers Beck and Cowan (2006), who studied the systemic nature of consciousness development, used a spiraling double-helix topology, which they called "the spiral of development," to depict the dialectic pattern in which bio-psycho-social systems emerge as humans respond to external life conditions. The spiral of development revolves between the demands of one's need to adapt to one's environment and one's desire to adapt the environment to one's self. The resulting emergent systems manifest as structures of core values ("value memes") that serve to organize both individual and collective consciousness.

Spiral Dynamics' structure-stages bear a close resemblance to the structures of consciousness outlined by Jean Gebser (1984) in his study of human history. In tracing the historical development of nearly every major field of human undertaking (e.g., art, science, language, literature, and philosophy), Gebser discerned an unfolding pattern of transformation that includes five consecutive worldview structures: archaic, magic, mythical, mental–rational, and integral. Each successive structure is characterized by a novel relationship toward space and time, while earlier structures continue to operate even as new ones emerge.

The most comprehensive effort to integrate the evolutionary dimensions of self, culture, and nature into a comprehensive map of reality was Wilber's (1995) AQAL (all-quadrants, all-levels) model. It depicts development as proceeding in four parallel hierarchical strands, or quadrants (see Figure 9.1). The quadrants represent four irreducible ontological dimensions that all living phenomena possess—*intentional, behavioral, cultural,* and *social* (or *systems*). They correspond to the interior and exterior of individual and collective dimensions of being. The quadrants also represent four fundamental epistemological perspectives through which every living phenomenon can be perceived.

AQAL does more than just represent existing reality. By using rich ontological language to describe the evolving structures of consciousness and culture, it invites us to reflect on the shared mechanisms that underlie evolution in all strands of reality. Such reflection brings forth and illumines new integral awareness, thus contributing to the formation of the emergent integral structure of consciousness. In other words, AQAL is not just a static map of reality, but an enactive paradigm; it is not just a theory of inclusion, but a space where inclusion of many perspectives can be practiced. We will next use this enactive paradigm to explore the evolution of conceptions of "What is mathematics?"

Stages of Mathematics

In an earlier work, Davis (1996) traced the history of mathematics, and proposed five major eras, or mentalities, into which it may be divided: *oral,*

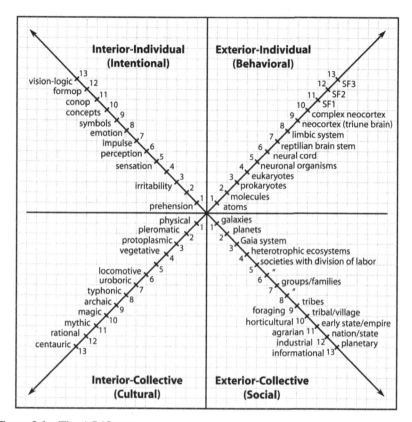

Figure 9.1 The AQAL map.

pre-formalist, formalist, hyper-formalist, and *post-formalist.* Integral philosophy suggests that these mentalities may form a dialectic sequence of increasingly complex human conceptions of mathematics, that is, a stage model for mathematics. We will show in this section that the five mentalities are developmental stages by correlating them with the broader structure-stages, or worldviews, used by integral writers (e.g., McIntosh, 2007; Wilber, 2006) to describe the development of consciousness. The worldviews are: *archaic, tribal, traditional, modernist, post-modern,* and *integral.* We will also show that the mentalities satisfy the subject–object mutuality that characterizes developmental processes, as indicated by Kegan (1994).

At each mentality, or stage, we explore these questions:

1. How does this stage respond to the question "What is mathematics?"
2. What is the connection between mathematics and the natural world?
3. How does this stage position the relationship between knowledge and knower?

4. What mathematical technologies are used at this stage?
5. How are mathematical truths validated?

The Oral Stage

This stage refers to societies that existed before the invention of writing in different parts of the world. In oral cultures, mathematics and mathematical meanings are found only in immediate experience and practical action. Mathematics is tightly bound to the knower's immediate environment in the natural world. Mathematical objects are classified by practical situation; numbers, for instance, are used as adjectives, rather than as nouns. Mathematical knowledge manifests in human processes, such as counting. The main technology of mathematics at this stage is oral narrative. The oral stage corresponds to the tribal stage of consciousness, in which tribal myths and immediate experience establish validity and truth.

The Pre-Formalist Stage

The invention of writing brought mathematics into the visual–representational realm. In the process, mathematical knowledge was preserved in symbol systems that enabled it to be understood over time and distance, and so it attained a similar separation from its knowers as the textual inscriptions of natural languages. At this stage, mathematics is understood as a mode of reasoning about unchanging forms, or essences, in the natural world. Mathematical knowledge resides outside of the knower, and is discovered by empirical observation. The technology of writing accords mathematics an independent existence through abstract objects, such as numbers, abstract categories, and geometric forms. This stage corresponds to traditional consciousness, in which scriptures that express the universal mythic order are the standards of validity and truth.

The Formalist Stage

The formalist stage of mathematical history originally emerged briefly in Ancient Greece, but it came into full fruition only in the early modern European era of Newton and Descartes. Mathematics is a distinct discipline in the formalist stage, with a separate body of knowledge and knowledge-producing methodology. The methodology of formal logic applies strict derivation rules to fundamental propositions, or axioms, in order to produce new mathematical results. At this stage, mathematical axioms, such as Euclid's Postulates, are framed in terms of observation from the natural world, though the description of the natural world is only one of several concerns of mathematics. The technologies of the formalist stage include the mechanism of formal proof, calculating devices, and formal mathematical representations, such as the Cartesian plane. This stage corresponds to modernist consciousness, in which reason is the standard of validity and truth.

The Hyper-Formalist Stage

The hyper-formalist stage arrived at the beginning of the 20th century, when the emergence of non-Euclidean geometries enabled mathematicians to manipulate the axioms of geometry. Mathematicians of the time, such as Hilbert and Russell, set out to reconstruct mathematics as a purely formal system with little or no correspondence to the natural world. Mathematical knowledge exists at this stage only to the extent that it can be derived within the logical parameters of a given formal system. The technologies of the hyper-formalist stage include non-standard logics and abstract grammars. The hyper-formalist stage is, in a sense, an extreme extension of formalist consciousness, as truth and validity are established solely by syntactic adherence to the rules of invented formal systems.

The Post-Formalist Stage

In the 1920s, Godel's proof of the incompleteness of formal systems challenged the hyper-formalist project, and cast mathematical certainty into doubt. The post-formalist stage regards mathematics as a socially-constructed interpretive discourse, rooted in our need to make sense of our environments and to construct our reality. Far from being separate from knowers, mathematical knowledge at this stage is embodied and enacted by both individual and collective knowers. The principal interpretive technology of this stage is deconstruction. This stage corresponds to the post-modern stage of consciousness, in which notions of validity and truth are themselves taken to be social and discursive constructions.

In Table 9.1, we list the subject and object at each of the five stages described thus far. The subjects respond to the question "What is mathematics?", and the objects address the question "What tools does mathematics use?"

As can be seen in Table 9.1, the stages follow a pattern of subject-object mutuality, in which the subject of one stage of development becomes the object of the subject of the next stage of development. At the oral stage,

TABLE 9.1   Subject–Object Mutuality in Davis' Stages of Mathematics

Stage	Subject (The nature of mathematics)	Object (The tools of mathematics)
Oral	Human processes and interactions about immediate phenomena	Objects in the environment
Pre-formalist	A mode of reasoning about essences	Abstract mathematical objects
Formalist	A separate discipline with a formal mode of reasoning	Rationality
Hyper-formalist	A formal system	Formalism
Post-formalist	A socially-constructed interpretive discourse	Discourse

mathematics consists of immediate experiences and human processes, such as counting. At the pre-formalist stage, the results of these processes are reified into abstract mathematical objects, such as numbers. Mathematics at this stage is a mode of reasoning that uses abstract objects to make sense of observed phenomena in the universe. At the formalist stage, the mode of reasoning itself is formalized into the mental construct that we know as "rationality." Mathematics at this stage is the body of knowledge derivable by formal proof from initial axioms drawn from the natural world. At the hyper-formalist stage, the process of formalism itself is objectified, and formal proof is seen as one of many possible formal logical systems. Mathematics at this stage consists of all results that can be derived by applying constructed logical rules to arbitrary initial axioms. Finally, at the post-formalist stage, both rationality and formalism are treated as discursive forms within a social matrix. Mathematics at this stage becomes a socially-constructed interpretive discourse.

The subject–object mutuality observed over the history of mathematics appears to confirm that the different mentalities represent a coherent evolution of conceptions of mathematics. This evolution follows a dialectic pattern, and proceeds through distinct developmental stages. We will now proceed to use this evolutionary understanding of mathematics to examine the current prevailing conceptions of mathematics, and their impact on mathematics pedagogy.

Current Conceptions of Mathematics

Much of mathematics education practice today seems to reside in the formalist and pre-formalist conceptions of mathematics. Many mathematics educators often regard mathematics as a static body of knowledge that represents extra-human reality. As Ernest (1985) noted, this Platonic view of mathematics greatly constrains mathematics pedagogy.

> By locating the source of mathematics in a pre-existing static structure, Platonism results in a static body-of-knowledge view of mathematics. Platonism discounts both man as a creator of mathematics and the importance of dynamic processes in mathematics. In educational terms this corresponds with the view of mathematics as an inert body of knowledge which instruction transmits to the student. (p. 607)

Ernest refers to an important connection between conceptions of mathematics and pedagogy. When mathematics is viewed as transcendent and essential, it follows that the teacher's role is to be a faithful conduit and gatekeeper for the established knowledge that informs the subject. This view supports a transmissive pedagogical model, which measures success according to conformity with pre-determined results, and employs systems of evaluation and discipline that aim for methodological conformity. Ernest argues against this model from a post-formalist perspective that values the enacted, creative, and dynamic human dimensions of mathematics.

Ernest is not alone among mathematics education researchers in criticizing transmission pedagogy that is founded on Platonic conceptions of mathematics. In the past three decades, a significant number of researchers have promoted what we are calling a post-formalist perspective of mathematics in education. They include mathematicians (e.g., Davis & Hersh, 1981) who reframed mathematics as one of the humanities, philosophers of mathematics (e.g., Lakatos, 1976) who analyzed the evolutionary dynamics implicit in mathematics production, and mathematics educators (e.g., Applebaum, 1995) who drew on cultural studies to resituate mathematics in social, ethical, and ideological terms. While these researchers have been instrumental in unearthing and elucidating some of the problematics of traditional and modernist education, postmodern thinking has gained little traction in the actual practice of mathematics education. Among all subjects in the school curriculum, mathematics seems most resistant to postmodern discourses of diversity and intersubjectivity.

From an evolutionary perspective, the problem is one of stagnation, or arrested development. Even though there is no shortage of ideas about the directions in which mathematical pedagogy might evolve, most mathematics educators remain largely indifferent to innovation. Notions of pragmatics, diversity, subjectivity, and intersubjectivity inevitably come up against the limiting Platonic conceptions of mathematics, and must contend with the certitude of "2 + 2 = 4." We now proceed to explore the question "What are some of the barriers to transcending Platonism?"

Barriers to Evolution

Modernist consciousness is characterized by rationality, and relies on the scientific method and objective reasoning for truth validation. For these reasons, mathematics has become the paramount technology of modernity. Without the tools of mathematics, neither physics, nor computers, nor stock markets would exist. The rapid pace of technological development in modern times would not have been possible without the time-honored truths of arithmetic, algebra, and calculus. Nowadays, mathematics plays an increasingly vital role in areas such as genetics, neuroscience, and ecology. Its significance is so ascendant in the information age that Baker (2008) has proposed that the *numerati* have now overtaken the *literati* in the role of defining cultural possibility.

It is important to note that the advances of mathematics in modern times were achieved with a Platonic perspective in mind. In other words, Platonic mathematics has served humanity well for many centuries. Given this complex interplay of intellectual inertia, economic investment, and emotional commitment, it is not surprising that there is massive resistance to interrogating, much less transforming the Platonic narrative. We often encounter individuals who argue passionately that mathematical truths, such as "2 + 2 = 4," are as real in our world as material objects, such as trees. Indeed, the Platonist position holds that mathematical truths are "more real" than trees, since they represent ideal forms that transcend material existence. Modernist consciousness tends to regard mathematics as the guidebook to the phenomena of the physical world. This prevailing view of mathematics is not likely to change, given the spectacular success of the present utilitarian conjunction between mathematics and scientific progress.

Educators are generally better positioned than the public at large to appreciate the shortcomings of the Platonic view of mathematics, as the defects of present-day school mathematics can often be traced to educational purposes that are supported philosophically by this view. It is strangely easy to argue that school mathematics is moribund. The deficiencies of the modernist–industrial model of education, which seeks to transmit preestablished results of mathematics, are becoming more and more apparent. Policy decisions of the last decade, requiring even more testing and even closer scrutiny of test results, have done little to enhance the experience of mathematics teaching and learning. On the contrary, the emphasis on hyper-instrumentality in mathematics education has likely contributed to increased student disaffection with the subject matter of mathematics. Yet, since educators operate within cultural and social structures that give rise to modern mathematics curricula, they are more likely to seek pragmatic instructional solutions than to deconstruct and revise their views on the subject matter. In our work, we have encountered many teachers who consider the question "What is mathematics?" as having little or no relevance to their practice.

Educators who wish to consider alternatives to Platonic assumptions about mathematics have to contend with certain cognitive barriers. For example, post-formalist perspectives are concerned with issues of intersubjectivity, justice, equity, class, race, and gender in mathematics. As Table 9.1 shows, the objects of post-formalist mathematics are discursive formations; they are very different from the mathematical objects of previous stages—numbers, formal proofs, and abstract grammars. This qualitative difference manifests in the language of mathematics education practice, and is very difficult to overcome. We regularly hear complaints from our pre-service teacher education students that our elaborations of post-formalist perspectives are "not mathematics," and that they belong, instead, to the disciplines of the social sciences and the humanities. When we speak with practicing teachers, we often sense that our post-formalist stance is removed from their daily experiences of prescribed learning outcomes, tests, marks, and correct answers.

Transcending Platonic assumptions about the nature of mathematical truth and ontology requires not only a profound change in language, but also a negotiation of some very difficult philosophical questions: *Is mathematics discovered or invented? What is the role of formal mathematics? What is truth? Can 2 + 2 equal anything other than 4? If so, what are the implications for our understanding of the natural world?* The absence of clear answers to these questions makes the certitude of Platonic mathematics all the more appealing to the average teacher.

Two accounts from the history of mathematics illustrate the staying power of the assumptions of Platonism in education. The New Math reforms of the 1960's sought to introduce a hyper-formalist perspective through a curriculum that included set theory, number bases other than 10, and Boolean algebras. New Math challenged the long-standing correspondence between school mathematics and phenomena in the natural world. In brief, these reforms failed within a few years, largely in response to complaints that this approach to mathematics was removed from the students' everyday experiences, and that many teachers and parents did not understand it fully. In 1989, the publication of NCTM's Standards heralded the introduction of an equity agenda that called for "mathematics for all." From an epistemological standpoint, the Standards challenged the traditional view of mathematics as a fixed body of knowledge by focusing on four strands: mathematics as problem solving, mathematics as communication, mathematics as reasoning, and mathematical connections. The implementation of reforms, which were based on the Standards, in California in the mid-1990's led to full-scale "math wars" (cf. Schoenfeld, 2004) that divided the educational community across the United States for the better part of a decade.

Integral philosophers (e.g., Wilber, 2006) assert that new worldview systems emerge only when previous ones become insufficient for dealing with the problematics created by changing life conditions. As Dewey (1910) similarly said of the transition from one worldview to another, "We do not solve them, we get over them" (p. 18). A critical mass of disorienting events or cognitive dissonances is required to move thinking to a new vantage point from which new aspects of reality can be seen. From our earlier discussion, it appears that the critical mass required to transcend Platonism is still building toward transition. But the impetus behind this research volume, titled *Unpacking Pedagogy: New Perspectives for Mathematics*, indicates the growing interest in alternatives to Platonic mathematics. Yet, it strikes us that, given the immense benefits that Platonic mathematics has brought to humanity, opposing the material achievements that are commonly linked to it would be counter-productive to the project of evolving mathematics. An approach that values and integrates the different stages is likely needed. We will now proceed to explore what such an approach might entail.

The Integral Perspective

Academic turf battles in mathematics education are the result of the attempts of the traditional, modernist, and postmodern consciousnesses to assert their perspectives in the field. The New Math reforms and the "math wars" of the 1990s are two examples of such battles. As we have seen, when it comes to perspectives on subject matter, the two main camps consist of those who view mathematics as transcendent and stable, and those who view it as embodied and emergent. While educators (Ball et al., 2005) agree that the disputes within the discipline do not serve the needs of students, there is little agreement on how to reconcile the seemingly opposed claims of the different camps within mathematics education. Our preceding analysis of the evolution of mathematics, combined with insights of integral philosophy, may offer one starting point for reconciliation.

Integral thinking highlights the importance of the evolutions of consciousness and culture to global well-being. Integral writers (e.g., Gebser, 1984; Wilber, 2006) pointed to the emergence of a new stage of consciousness, or epistemological interpretive framework, which they called "integral consciousness." In contrast to mental–rational consciousness that precedes it, integral consciousness is multi-perspectival, that is, characterized by lack of attachment to monological perspectives. Since all preceding structures of consciousness in the spiral are transparent to integral awareness, it is able to integrate them and "live through" them, rather than be controlled by any one of them. While the integral structure does not come with a ready-made set of values, it seeks to solve problems by bringing as many perspectives as possible to bear on given situations in order to arrive at appropriate, yet never permanent, solutions.

Integral consciousness seeks to promote the health of the entire evolutionary spiral by acknowledging the dignities and contributions of every significant historical worldview—including the traditional, modernist, and postmodern. Likewise, it seeks to overcome constructively the limitations of these worldviews. Doing so requires the emergent capacity of *vision-logic*, which Kegan (1994) described as "the capacity to see conflict as a signal of our overidentification with a single system" (p. 351). Vision-logic is network logic. It is the ability to see how different elements of an evolutionary system work together dialectically, to "live through" their different perspectives, and to harmonize them by valuing each element for its contribution to the entire system.

From an integral perspective, the question of whether mathematics is Platonic or embodied, stable or emergent, is founded on false dichotomies. Mathematics is all of these things and more. Every conception of mathematics has developed in response to a different set of life conditions, and its values and practices have their appropriate applications under these life conditions. The evolution of new conceptions, or stages, does not negate earlier ones, but rather exhibits a dialectic pattern of transcendence and inclusion. In this pattern, new stages retain the robust structures of earlier stages, while "complexifying" them into higher-order unities. In other words, mathematics is both Platonic and embodied, stable and emergent, objective and socially-constructed, fixed and enacted, a science and a branch of the humanities. The observer's vantage point and context determine which aspect of mathematics is revealed at any given moment.

With this understanding in mind, we can see that formal mathematics is a narrative that captures and codifies the more stable dimensions of mathematics. It does so by turning human processes and thought patterns into abstract discursive objects that, over time, take on transcendent, universal qualities. In her analysis of the development of mathematical discourse, Sfard (2008) explained that the process of objectification is indispensible, as it provides mathematical discourse with its principal advantage—the ability to represent complex ideas with concision and compactness. By eliminating the temporal dimension of phenomena, objectification also helps users of mathematics cope with the fluidity of human experience.

From a post-formalist perspective, Platonic conceptions of mathematics often appear to be wrongheaded and even dangerous. Postmodern critiques (e.g., Appelbaum, 1995; Walshaw, 2004) have deconstructed Platonism, and identified various problems that arise when formal mathematics is treated as the only "real" mathematics. Among these problems are: marginalization of other mathematics, hegemony of Western thought, inequitable access to mathematics, and alienation of knowers. However, from an integral perspective, Platonism should be valued for its appropriate contributions in certain contexts, just as post-formalist perspectives ought to be honored for their appropriate contributions in other contexts. One should not seek to defeat Platonism, but rather to harmonize the stability that it has conferred on mathematics with emergent and subjective potentialities.

The difference between integral and post-formalist perspectives' views on Platonism signals that the integral perspective may be a new stage in the evolution of mathematics that will succeed the post-formalist stage. As discussed earlier, at the post-formalist stage, mathematics is a socially-constructed interpretive discourse. The integral stage organizes interpretive discourses into evolving bio-psycho-social systems. At this stage, mathematics becomes an evolving system of interpretive discourses, or perspectives. As Table 9.2 shows, Kegan's subject–object mutuality is again confirmed.

TABLE 9.2 Subject–Object Mutuality in the Latest Two Stages in the Evolution of Mathematics

Stage	Subject (The nature of mathematics)	Object (The tools of mathematics)
Post-formalist	A socially-constructed interpretive discourse	Discourse
Integral	An evolving system of interpretive discourses (perspectives)	Interpretive discourses

APPLYING IDEAS

Technologies of Mathematics

The integral perspective invites us to harmonize the Platonic conception of mathematics with emergent and embodied mathematics. As we embark on this project of integration, we accept that the practice of mathematics pedagogy has so far favored Platonism almost exclusively over emergence–embodiment. So any intervention that would ease pedagogy's pervasive bias toward Platonism is welcome.

We have searched for such interventions in our work with pre-service teachers. At first, we opted for a philosophical approach, and asked our students to consider the question "Is mathematics discovered or created?" Most of them appeared to be uninterested in this highly theoretical question, and even those who attempted to respond were soon tangled in abstractions. We concluded that concrete examples were needed to encourage meaningful debate around the topic.

As we tried out different mathematical examples, we discovered that most of them led back to Platonic identifications held by the pre-service teachers; only a few examples prompted more emergent reactions. For instance, no teacher in the class was willing to believe that the equation $2 + 2 = 4$ is a human construction. The students argued, with considerable conviction and passion, that this equation represents a universal truth that holds true for everyone, everywhere, and for all time. In fact, they argued, it would still hold true if there were no knowers in the universe to know it. On the other hand, when we introduced the abstract operator "↑" as $a \uparrow b = 2a - b$, the students were unanimous in their agreement that the equation $3 \uparrow 2 = 4$ is a human construction. They explained that the abstract operator "↑" had just been defined by us, and so it must be a human creation. When we suggested that the operator "+" was also defined by humans at some point in history, some students responded that "+" is a "real" operator, while "↑" is not.

In general, we found that our students showed an emotional commitment to the transcendence and timelessness of the results of formal mathematics. The more elementary the result, the more committed students were to its permanent status. Wilber (2006) referred to significations that have such transcendent status as *Kosmic habits*; in Foucauldian language, they are called *technologies*. The older the habit or technology, the more self-evident and secure it is to those who participate in it. For example, while the rudimentary equation $2 + 2 = 4$ was created by social agreement at some early point in human history, once it was formalized into language and conventional signs, it acquired an independent and objective existence beyond a series of historically determined symbols. With every new generation, the reality of $2 + 2 = 4$ became more entrenched and transcendent. Some may go so far as to say that it has an ontological status apart from actual objects and significations within language. Likewise, many concepts of school mathematics have been in use for so long that they have attained fixed meanings that conceal the circumstances that gave rise to them over time. Yet, teaching and learning of mathematics are, in fact, heavily dependent on multiple meanings that emerge through interpretation. As fixed as $2 + 2 = 4$ may seem to us, as determiners of its metaphysical status, every pedagogical encounter enhances the collective meaning of $2 + 2 = 4$ by occasioning unique subjective meanings.

In order to reveal the constructed nature of mathematical technologies, we continued searching for examples of interpreted objects from school mathematics that are not as contrived as the abstract operator "↑". In our search, we identified coordinate geometry as a useful example. Many of our students readily agreed that the invention of the coordinate plane by Descrates in 1637 allowed mathematicians to conceive of geometric figures, such as circles, in a way that is radically different than that used by Greek geometers. We then probed further by asking, "What are circles?" Some of the teachers began to appreciate that, rather than being pre-existing objects, circles are created through human interpretive frames.

Logarithms turned out to be another good example, as the students realized that they were invented by Napier in 1614 as a technology for multiplying large numbers. We found that our pre-service teachers preferred the language of "technologies of mathematics" to that of "emergent mathematics." We suppose that their preference is attributable to their familiarity with the metaphors of technological development, and to persistent Platonic commitments to original meanings.

Coordinate geometry and logarithms proved to be productive examples for our pre-service teachers to consider because, from a historical perspective, the emergence of these concepts has provided new metaphors for pre-existing mathematical objects. As our experiences with the pre-service teachers indicate, the emergent–embodied nature of mathematics can be disclosed by engaging concrete and familiar mathematical examples that lend themselves to interpretation through multiple images and metaphors. We will now proceed to describe an ongoing study that we have been conducting with a group of experienced teachers around the meanings of multiplication.

Emergent–Embodied Mathematics in Action

Our study has unfolded over a period of two years. It involved a group of 11 experienced middle-school teachers who gathered in monthly concept-study meetings to discuss and deconstruct different curriculum topics. These ongoing meetings have been conceived as collective knowledge-producing occasions, through which mathematics educators identify, interpret, interrogate, invent, and elaborate images, metaphors, analogies, examples, exemplars, exercises, gestures, and applications that are invoked in efforts to support the development of students' mathematical understandings. The concept of multiplication has received the most attention from the group. Indeed, discussions of other concepts have regularly gravitated to the subject of multiplication.

We began with the direct question "What is multiplication?" After the two most obvious answers—"repeated addition" and "grouping"—were given, we asked, "And what else?" The rest of the morning was organized around discussions of pedagogical difficulties that arise in teaching multiplication, investigations of when and how elaborations are introduced, and analyses of teaching resources for multiplication. The end result was a listing of metaphors, images, analogies, and applications, as shown in Figure 9.2.

Multiplication involves...
- repeated grouping
- repeated addition
- sequential folding
- layering
- the basis of proportional reasoning
- grid-generating
- dimension-changing
- intermediary of adding and exponentiating
- opposite/inverse of division
- stretching or compressing of number-line
- magnification
- branching
- rotating a number line
- linear function
- scaling
- and so on...

Figure 9.2 A teacher-generated list of interpretations of multiplication.

The teachers seemed surprised at the lengthy list of realizations of the signifier *multiplication* that they were able to generate. Their surprise was summarized by the comment: "Apparently, we don't have a good handle on what we know yet." The teachers also recognized that the point of the list was not to provide an exhaustive summary of interpretations of multiplication, but rather to indicate the range of associations that were accessible to the group on this day.

The next major development on the subject of multiplication took place several months later, when a few of the participants urged the group to organize all of the realizations in the list according to grade levels and thematic categories. The resulting chart is shown in Figure 9.3.

Upon examining their mapping, participants were surprised to realize that distinct and coherent strands of interpretation are systematically developed over the K–12 experience. The different realizations of multiplication, far from being random or isolated, were organized into grander interpretive structures. The teachers acknowledged that they had, in fact, participated in the systematic development of these structures prior to the study, without having reflected on their participation.

The next development took place when the group tackled the question "Is 1 prime?" The teachers began to explore the relevance and implications of different realizations of multiplication to the question. In the discussion, participants often framed their remarks as "If..., then..." statements, locating their comments within specific metaphorical domains. In other

GRADE LEVEL	APPLICATIONS/ ALGORITHMS	ARITHMETIC INTERPRETATIONS		PARTITIONAL INTERPRETATIONS	COMPOSITIONAL INTERPRETATIONS
		Based on Sets of Objects	Based on Lines		
12					
	vectors				
11	matrices		scaling/slope (a function)		
10					
9	polynomials irrationals				
8	integers		numberline stretching/compressing (& rotating)		
					dimension-jumping*
7	common fractions				
6		proportional reasoning			area-producing ("by")
	decimal fractions			folding*	
5			numberline hopping	splitting*	
4	multidigit wholes	repeated addition ("times")			
3				branching*	
	wholes	array-making			
2		grouping ("of")			
1					
K					* signifies a unified definition of 'factor'

Figure 9.3   An evolving landscape of the concept of multiplication.

If multiplication is:	... then a product is:	... a factor is:	... a prime is:	Is 1 prime?
REPEATED ADDITION	Sum (e.g., $2 \times 3 = 2 + 2 + 2 = 3 + 3$)	either an addend or a count of addends	a product that is either a sum of 1's or itself.	NO – 1 cannot be produced by *repeatedly* adding any whole number to itself.
GROUPING	a set of sets (e.g., 2×3 means either 2 sets of 3 items or 3 sets of two items	either the number of items in a set, or the number of sets	a product that can only be made when one of the factors is 1	YES – 1 is one set of one.
BRANCHING	the number of end tips on a 'tree' produced by a sequence of branchings (e.g., 2×3 means ⌄⌄ ⌄⌄ ⌄⌄ ⌄)	a branching (i.e., to multiple by n, each tip is branched n times)	a tree that can only be produced directly (i.e., not as a combination of branchings)	NO – 1 is a starting place/point ... a pre-product as it were.
FOLDING	number of discrete regions produced by a series of folds (e.g., 2×3 means do a 2-fold, then a 3-fold, giving regions)	a fold (i.e., to multiply by n, the object is folded in n equal-sized regions using $n–1$ creases)	a number of regions that can only be folded directly	NO – no folds are involved in generating 1 region
ARRAY-MAKING	cells in an m by n array	a dimension	a product that can only be constructed with a unit dimension	YES – an array with one cell must have a unit dimension

Figure 9.4 Some analogical implications of different realizations of multiplication.

words, participants were consciously engaging in analogical, as opposed to logical, reasoning. Some of the results are presented in Figure 9.4.

The teachers came to appreciate through first-hand experience that humans are not merely logical creatures, but association-making beings whose capacity for formal reason operates alongside their predisposition to make connections (cf. Lakoff & Johnson, 1999). This appreciation became the central point of the engagement, so much so that one of teachers remarked, "No wonder the kids find that so difficult."

A recent layer in the group's ongoing concept study of multiplication is that of shared reconciliation of seemingly different realizations of multiplication into unified blends. The first blend was a grid-based representation of multiplication, shown in Figure 9.5, which pulls together several realizations—including repeated addition, array-making, and area-making—as it highlights procedural similarities in handling additive multiplicands across diverse number systems and algebraic applications.

Another blend was created when a teacher noted that the number-line-stretching interpretation could be combined with the mapping-function interpretation, as shown in Figure 9.6. This blend led directly to an intuitive graphical "proof" of the result that the product of two negative numbers must be positive. The teachers were very satisfied with their efforts in creating new mathematics, which had not been previously encountered by anyone in the room.

By opening up a familiar mathematical concept for hermeneutic questioning and elaboration, our concept study of multiplication underscored the dynamic, embodied, and enacted dimensions of mathematical knowledge. As the teachers explored tacit layers of mathematical knowledge,

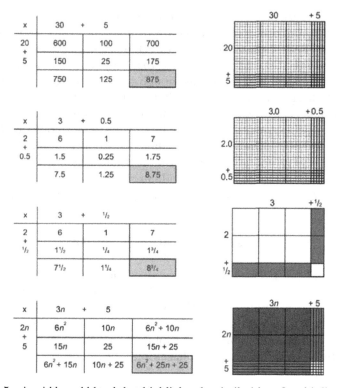

x	30	+	5
20 +	600	100	700
5	150	25	175
	750	125	875

x	3	+	0.5
2 +	6	1	7
0.5	1.5	0.25	1.75
	7.5	1.25	8.75

x	3	+	$\frac{1}{2}$
2 +	6	1	7
$\frac{1}{2}$	$1\frac{1}{2}$	$\frac{1}{4}$	$1\frac{3}{4}$
	$7\frac{1}{2}$	$1\frac{1}{4}$	$8\frac{3}{4}$

x	$3n$	+	5
$2n$ +	$6n^2$	$10n$	$6n^2 + 10n$
5	$15n$	25	$15n + 25$
	$6n^2 + 15n$	$10n + 25$	$6n^2 + 25n + 25$

Figure 9.5 A grid-based blend that highlights the similarities of multiplicative processes involving additive multiplicands.

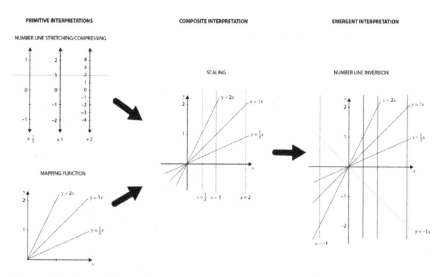

Figure 9.6 A graph-based blend that combines linear models of multiplication.

and as they constructed emergent knowledge through collaboration, their Platonic assumptions about the nature of mathematical knowledge were challenged. Reflecting on the group's extended engagement with multiplication, some of the participants said that they "have really been able to get inside the idea" and to "feel as though [they're] really contributing to how multiplication is understood." Their understandings of multiplication stood in sharp contrast with their general conception of mathematics. As two of the teachers put it, "It feels like it's outside of us" and "... we feel like we're outside of it." The teachers felt alienated from (Platonic) mathematics, yet inhabiting, and even responsible for, (participatory) multiplication. The contrast reflects the need for mathematics educators to transcend Platonism, and to develop a more participatory and generative pedagogy.

LIVING MATHEMATICS PEDAGOGY

The integral perspective, which recognizes the systemic evolutionary dimensions of mathematics, calls on teachers to embody and enact a living pedagogy that promotes the evolution of culture and consciousness. To achieve this goal, educators should be aware of the structures of evolution in their field, pay attention to the evolutionary tensions among different perspectives on mathematics, and harmonize these tensions as they arise in pedagogical moments.

As discussed previously, the evolution of mathematics is governed by a pronounced tension between stability and novelty. Stability manifests in conceptions of mathematics as Platonic, fixed, explicit, and formal. Novelty manifests in conceptions of mathematics as emergent, embodied, tacit, enacted, and participatory. The stable, transcendent conception of mathematics dominates mathematics teaching and learning at the moment. Therefore, educators who wish to promote cultural evolution in mathematics need to find and employ skillful means for transcending Platonism. In our work, we have identified several principles that may assist educators in this task. They are: elaborate the specific, encourage multiplicity, and use active language. We will examine each of these principles in turn.

As the concept study of multiplication has shown, elaborative engagement with specific mathematical concepts can be an effective way to uncover emergent–embodied features of mathematics. Our experience indicates that the more basic the mathematical concept under scrutiny, the more confidently participants engage with it. The process of elaboration reduces expectations for performative instrumentality, which are prevalent in current mathematics pedagogy. The focus of mathematical inquiry then shifts from the question "What is the correct answer?" to the question "What else is there to say about the mathematical situation?" The shift takes us from exclusive and fixed mathematical meanings to shared and subjective meanings. The activity of hermeneutic elaboration enables participants to enact, reflect on, and explicate tacit and embodied knowings. In the process, participatory learning environments that can accommodate and encourage multiple opinions are created.

Multiplicity is a crucial element of living pedagogy, since the "bumping together" of different ideas is essential for the emergence of novelty (cf. Davis and Sumara, 2006). Mathematics educators can promote multiplicity in teaching and learning in a variety of ways. Some of them are: employing a multiplicity of mathematical metaphors and images to understand a mathematical concept, assessing a multiplicity of approaches to solving a given problem, and engaging a multiplicity of curricular and non-curricular topics in instruction. It is hoped that students who are exposed to multiplicity and plurality in classroom discourse will be open to the multiple meanings and different interpretation of mathematical concepts.

The language that educators choose to use may either enable or constrain students' interpretive possibilities. For example, the instructions "Find the correct answer" and "Think of different ways in which you may approach this mathematical situation" may initiate different learning processes. Platonic mathematics is characterized by a comprehensive, fully-alienated discourse about mathematical objects. Educators who would like to approach mathematics differently should strive to alter their discursive practices by replacing some references to objects with descriptions of human actions. For example, a teacher may define a triangle by saying: "A triangle is a 3-sided polygon," or alternatively by saying, "We shall call any 3-sided polygon a triangle." The latter utterance reminds learners that the human act of naming is co-implicated in every mathematical definition. Admittedly, altering the well-rehearsed alienated discourse of Platonic mathematics is no small task. However, engaging in this project can sharpen pedagogues' sensitivities to the subtle ways in which discourse shapes conceptions of mathematics in their classes.

Educators who adopt the three principles—elaborate the specific, encourage multiplicity, and use active language—will be providing some of the necessary conditions for emergence of novelty (cf. Davis & Sumara, 2006) and dialogue (cf. Sidorkin, 1999) in their classes. These educators should be ready to receive the new mathematical meanings that are sure to appear. Students today are exposed to vast amounts of mathematical information through television, video games, and the Internet. This information provides numerous living contexts for novel constructions of mathematical meanings. When these contexts come up in classroom interactions, teachers ought to take interest in them, since they are the learners' true lived experience. Teachers should be willing to investigate these contexts together with their students, explicate the mathematics that emerges, and connect this new mathematics to pre-existing mathematical results.

An integral living pedagogy offers educators the opportunity to teach an ever-evolving and innovative subject matter. Teachers who are aware of the dialectic processes of evolution may negotiate the rich stability of formal mathematics with the refreshing novelty of their students' emergent interpretations. Teachers and students co-create mathematics by elaborating, deconstructing, and infusing mathematical concepts with new layers of meaning. The integral perspective views both pre-established and emergent contexts, the past and the present, as valuable sources of inspiration for hermeneutic elaboration.

Living mathematics pedagogy situates teachers as vital participants in the creation of mathematical possibilities. Far from being peripheral agents who passively transmit established results of mathematics, teachers give shape and substance to cultural mathematics. An integral pedagogy of life stands to breathe new life into school mathematics and cultural mathematics, just as Edgardo Cheb-Terrab's computer tools breathe new life into long-standing problems of research mathematics.

REFERENCES

Appelbaum, P. (1995). *Popular culture, educational discourse, and mathematics.* Albany, NY: State University of New York Press.

Baker, S. (2008). *The numerati.* Boston: Houghton Mifflin.

Ball, D. L., Ferrini-Mundy, J., Kilpatrick, J., Milgram, J., Schmid, W., & Schaar, R. (2005). Reaching for common ground in K–12 mathematics education [Electronic version]. *Notices of the American Mathematical Society, 52*(9), 1055–1058.

Cheb-Terrab, E. S, & Roche, A. D. (2003). An Abel ordinary differential equation class generalizing known integrable classes, *European Journal of Applied Mathematics, 14*(2), 217–229.

Baldwin, J. M. (2007). *Thought and things: A study of the development and meaning of thought, or genetic logic.* Whitefish, MT: Kessinger Publishing. (Original work published 1906)

Beck, D. E., & Cowan, C. C. (2006). *Spiral dynamics: Mastering values, leadership, and change.* Malden, MA: Blackwell Publishing.

Davis, B. (1996). *Teaching mathematics: Toward a sound alternative.* New York: Garland.

Davis, B., & Sumara, D. (2006). *Complexity and education: Inquiries into learning, teaching, and research.* Mahwah, NJ: Erlbaum.

Davis, P. J., & Hersh, R. (1981). *The mathematical experience.* Boston: Houghton Mifflin.

Dewey, J. (1910). *The influence of Darwin on philosophy (and other essays).* New York: Henry Holt.

Ernest, P. (1985). The philosophy of mathematics and mathematics education. *International Journal for Mathematics Education in Science and Technology, 16*(5), 603–612.

Gebser, J. (1984). *The ever-present origin* (Trans: N. Barstad & A. Mikunas). Athens, OH: Ohio University Press. (Original work published 1949).

Graves, C. W. (1970). Levels of existence: An open system theory of values. *Journal of Humanistic Psychology, 10*(2), 131–155.

Kegan, R. (1994). *In over our heads: The mental demands of modern life.* Cambridge, MA: Harvard University Press.

Lakatos, I. (1976). *Proofs and refutations: The logic of mathematical discovery.* Cambridge, UK: Cambridge University Press.

Lakoff, G., & Johnson, M. (1999). *Philosophy in the flesh: The embodied mind and its challenge to Western thought.* New York: Basic Books.

McIntosh, S. (2007). *Integral consciousness and the future of evolution: How the integral worldview is transforming politics, culture and spirituality.* St. Paul, MN: Paragon House.

Schoenfeld, A. (2004). The math wars. *Educational Policy, 18*(1), 253–286.

Sfard, A. (2008). *Thinking as communicating: Human development, the growth of discourses, and mathematizing.* New York: Cambridge University Press.

Sidorkin, A.M. (1999). *Beyond discourse: Education, the self, and dialogue.* Albany, NY: SUNY Press.

Walshaw, M. (Ed.). (2004). *Mathematics education within the postmodern.* Greenwich, CT: Information Age.

Wilber, K. (1995). *Sex, ecology, spirituality: The spirit of evolution.* Boston: Shambhala.

Wilber, K. (2006). *Integral spirituality: A startling new role for religion in the modern and postmodern world.* Boston: Integral Books.

FURTHER READING

Readers who are interested in evolutionary dimensions of mathematics and mathematics education may enjoy some additional readings on integral philosophy and complexity science. Wilber (1995) is the seminal text that introduced the AQAL map, and is a good starting point for exploring integral philosophy. Davis and Sumara (2006) provide a detailed study of the contributions of complexity thinking to educational research and practice. Sfard (2008) describes the mechanisms of the evolution of mathematical discourse.

CHAPTER 10

DECONSTRUCTING DISCOURSES IN A MATHEMATICS EDUCATION COURSE

Teachers Reflecting *Differently*

David Stinson and Ginny Powell

ABSTRACT

What should a mathematics classroom look like? And what should mathematics teachers do? Over the past 30 years, the National Council of Teachers of Mathematics (NCTM) has produced a collection of documents that provide suggestions for these politically charged questions (see NCTM, 1989, 1991, 1995, 2000; Kilpatrick, Martin, & Schifter, 2003). Most mathematics teachers, educators, and policymakers would agree that these documents describe a different mathematics classroom than that which is experienced by most students in schools in the USA, and throughout schools in English-speaking countries (see Hiebert, 2003, for a discussion of "traditional" curricula and pedagogy). It has also been argued that these documents are the catalyst for the "math wars" armed by two opposing camps: the traditionalists and the

Unpacking Pedagogy: New Perspectives for Mathematics Classrooms, pages 201–221
Copyright © 2010 by Information Age Publishing

reformers (Schoenfeld, 2004). The traditionalists' camp fears that the NCTM reform-oriented, "standards-based" curricula are superficial and undermine "classical" mathematical values, whereas the reformers' camp claims that the reform-oriented curricula reflect a deeper, richer view of mathematics (p. 253). The purpose of this chapter is to explore how postmodern theory might assist mathematics teachers (educators and policymakers) who position themselves in either camp (or somewhere in between) to begin to ask *different* questions regarding teachers and students of mathematics, and what it means to practice "good" mathematics teaching.

Although the impact of the NCTM documents in actually reforming mathematics teaching has been somewhat limited (see Wilson & Goldenberg, 1998), research has shown that these documents have had an impact on how mathematics teachers define and practice good mathematics teaching (Wilson, Cooney, & Stinson, 2005). Wilson, Cooney, and Stinson's research on the perspectives of seasoned mathematics teachers about good teaching suggests that efforts to reform mathematics teaching are seldom all-or-nothing affairs. Their research illustrated that even as seasoned teachers reformed (some of) their teaching practices, most often, they continued to maintain a belief in the teacher-centered classroom and the infallibility of mathematics. It has been argued that the latter of these beliefs is counter to reform-oriented mathematics teaching, thus securing the continuation of traditional practices (see Davis & Hersh, 1981; Ernest, 1998). To make it possible for teachers to create mathematics classrooms that are consistent with the constructivist, student-centered objectives of reform-oriented mathematics teaching, we believe that teachers must be provided an opportunity to challenge and "trouble" both traditional mathematics teaching and the reform efforts themselves.

Here, we broadly define reform-oriented education from a Deweyian experiential prospective (see Dewey, 1916, 1998), in that the aims of "effective" teaching must be an outgrowth of experience, provide flexibility to meet various circumstances, and represent a freeing of activities (Dewey, 1916). Specifically, within the context of mathematics education, we draw upon Davis and Hersh (1981), who stated: "Ideally, mathematical instruction says, 'Come, let us reason together'" (p. 282), and Ernest (1998), who reminded us that mathematics "for all its wonder...remains a set of human practices, grounded, like everything else, in the material world we inhabit" (p. xii). In understanding the mathematics classroom as a pedagogical space for teachers and students to reason together through the socially constructed discipline of mathematics, we believe that postmodern theory provides a different theoretical framework for teachers to trouble both traditional and reform-oriented mathematics teaching as they explore their own pedagogical philosophies and practices.

The value of postmodern theory is found in its awareness of and tolerance toward social differences, ambiguity, and conflict; it requires developing new languages, conventions, and skills to address the moral and political implications of knowledge (Seidman, 1994). In short, postmodern theory requires shifting the "focus from foundations and familiar struggles of establishing authority toward exploring tentativeness and developing scepticism of those principles and methods that put a positive gloss on fundamentals and certainties" (Walshaw, 2004b, pp. 3–4). Gordon (1980) marked the beginning of postmodern theory in the mid-to-late 1970s, characterizing the immediate years after the failed Student and Worker's Revolt of 1968 in Paris as an unusual, fascinating, and confused period in which new lines of investigation and critique emerged on the intellectual scene. Some of these new lines of investigation and critique can be found in Foucault's (1972) reinscription of *discourse*, Derrida's (1997) *deconstruction* of language and cultural practices, and Deleuze and Guattari's (1987) characterization of the *rhizome*.

INTRODUCING IDEAS

Discourse and Deconstruction, Rhizome and Reflection

As Foucault (1972) reinscribed the concept *discourse*, he argued that discourses are not a mere intersection of words and things, but are "practices that systematically form the objects of which they speak" (p. 49). That is to say, for Foucault, "discourses do not merely *reflect* or represent social entities and relationships; they actively construct or *constitute* them" (Walshaw, 2007, p. 19, emphasis in the original). Foucault (1990), however, also conceived discourses "as a series of discontinuous segments whose tactical function is neither uniform nor stable" (p. 100); therefore, we are not forever doomed by discourses. In joining power (which is reinscribed in postmodern theory, and discussed later in the chapter) and knowledge through discourse, Foucault identified discourse both as an effect of power and as providing points of resistance, noting that discourse not only transmits, produces, and reinforces power, but also undermines and exposes power, renders it fragile and makes it possible to obstruct it (p. 101). In other words, although Foucault claimed that discourses structure knowledge, their lack of uniformity and stability make discourses vulnerable to resistance, providing for the occasion of developing different discourses—and, in turn, different knowledges.

In general, Foucault's (1972) reinscription of discourses replaces the concept of the "nature" of knowledge with the "discursive formation" (p. 38) of knowledge. His reinscription rejects the "natural" or taken-for-granted concepts of knowledge found in humanism, such as Descartes' dualism of mind–body (which argues that the thinking subject is the authentic author of knowledge) or Comte's positivism (which argues for a "scientific" knowledge gained from methodologically observing the sensible universe) (St. Pierre, 2000). Specifically, Foucault uncovered knowledge as a discursive formation through the means of performing an archeological analysis. The methods of an archaeological analysis examine the history of discourses; but, rather than being concerned with uncovering the "truth" by an examination of facts and dates, it is concerned with the "historical conditions, assumptions, and power relations that allow certain statements, and by extension, certain discourses to appear" (St. Pierre, 2000, p. 496). This examination led Foucault to ask, "How is it that one particular statement appeared rather than another?" (1972, p. 27). Discursive formations (including knowledge), therefore, are open to reconfiguration because they are produced historically and culturally in relations of power (see Foucault, 1990). That is to say, there is no origin, or, understood in another way, no center to discourses (Derrida, 1978).

This lack of center leaves discursive formations open to what Derrida (1997) identified as *deconstruction*. Although Derrida refused to limit the possibilities of deconstruction through definition (see Derrida & Montefiore, 1992, 2001), others have described it as the methodology of exposing discursive binary oppositions defined interdependently by mutual exclusion, such as good/evil or true/false (Dillon, 1999). For Derrida, these binary oppositions attempt to shape the very structure of thought by constructing an "essential" center and authorizing presence—a center and presence that, it is assumed, will collapse if the binary opposition is undermined (Usher & Edwards, 1994). Within the context of mathematics education, some of these binary oppositions are: mathematical Truths/mathematical *t*ruths, teacher/student, effective teacher/non-effective teacher, traditional teaching/reform teaching, mathematically able student/non-mathematically able student, high-level course/low-level course, and so forth.

The deconstruction of binary oppositions identifies the first term, that is, the "privileged" term, as being dependent on its identity by the exclusion of the other term, demonstrating that primacy really belongs to the second term, that is, the subordinate term, instead (Sarup, 1993). In Derrida's deconstruction, it is not enough to just neutralize the binary oppositions, but one must reverse and displace the binaries. He argued that there is a violent hierarchy within binaries, in that one of the two terms controls the other, holding a superior position (Sarup, 1993).

Deconstruction then involves unsettling and displacing (or troubling) binary hierarchies, uncovering their historically contingent origin and politically charged roles, not to provide a "better" foundation for knowledge and society, but to dislodge their dominance, creating a social space that is tolerant of difference, ambiguity, and playful innovations that favors autonomy and democracy (Seidman, 1994). That is to say, deconstruction is not about tearing down, but about rebuilding by looking at how a structure has been constructed, what holds it together, and what it produces (St. Pierre, 2000). Or, said more directly, deconstruction acknowledges that the world has been constructed through language and cultural practices;

consequently, it can be deconstructed and reconstructed again and again (St. Pierre, 2000).

Deleuze and Guattari's (1987) characterization of the *rhizome* continues this deconstructive process in a non-linear (i.e., rhizomatic) manner. The *rhizome*, as described by Deleuze and Guattari, is not "reducible neither to the One nor the multiple.... has neither a beginning nor end, but always a middle (*milieu*) from which it grows and overspills" (p. 21). This concept permits not only a different direction to thought or meaning (i.e., discourses), but multiplicitous directions at once. Meaning, therefore, need never be settled upon, but suspended in a state of becoming, allowing constant change and multiple "lines of flight" (p. 9). The self, others, and knowledge can thus be seen as multiplicitous. From a practical standpoint, it is only when such multiplicitous meanings are present that one meaning can be chosen to act upon; without multiplicity, there is no choice.

Choosing among multiplicitous options is at the heart of Dewey's (1989) concept *reflective thinking* (a familiar concept for most readers). Within the past two decades or so, the discourse of *reflection* has become a standard that all teachers (and students) must strive toward, being identified as a crucial characteristic of exemplar teachers by numerous national, state, and local organizations, foundations, and boards (Rodgers, 2002). The dominance of the discourse of reflection has resulted in the proliferation of teacher reflection education literature. Present in this literature are essays that describe professional development models for mathematics teachers that position teacher reflection as foundational in teacher change (e.g., Hart, Najee-ullah, & Schultz, 2004), as well as research studies that identify teacher reflection as a key component in mathematics teacher action research (e.g., Jaworski, 1998; Thomas & Santiago, 2006).

Mewborn (1999), in her study on reflective thinking among pre-service elementary mathematics teachers, traced the emphasis of teacher reflection to Dewey's (1974) essay "The Relation of Theory to Practice in Education." Dewey believed that the primary purpose of teacher education should be to help teachers reflect on problems of practice, believing teachers who lack an inquiry mind (i.e., reflective thinking) will have their professional growth curtailed (Mewborn, 1999). Although Mewborn rightly noted that there is little agreement as to the content and nature of reflective thinking, in general, she did identify three common themes that are present within the literature: (1) reflective thinking is qualitatively different from recollection or rationalization, (2) action is an integral component of reflective thinking, and (3) reflective thinking is both an individual and shared experience. We add a fourth common theme, in that, similar to Mewborn, much of this literature traces the discourse of reflective thinking to John Dewey.

In "How We Think," Dewey (1989) explicitly defined reflective thought as "*active, persistent, and careful consideration of any belief or supposed form of knowledge in the light of the grounds that support it and the further conclusions to which it tends*" (p. 118, italics in original). Dewey elaborated on this definition, writing:

> We may carry our account further by noting that *reflective thinking*, in distinction from other operations to which we apply the name of thought, involves (1) a state of doubt, hesitation, perplexity, mental difficulty, in which thinking originates, and (2) an act of searching, hunting, inquiring, to find material that will resolve the doubt, settle and dispose of the perplexity. (pp. 120–121)

In other words, "one can think reflectively only when one is willing to endure suspense and to undergo the trouble of searching" (p. 124). Dewey identified five phases (or aspects) of reflective thought: (1) suggestion; (2) intellectualization; (3) guiding idea, hypothesis; (4) reasoning; and (5) testing the hypothesis by action (pp. 200–206).

Dewey (1989) claimed that the first phase not only involves a person encountering two or more ideas or suggestions (i.e., a problem), but also, and most important, requires maintaining a state of suspense before acting, producing further inquiry among competing suggestions to examine the differing conditions, resources, difficulties, and obstacles that might exist. The second phase requires that the person positions the problem within its social context, making note of one's present position, so as to distance oneself from the problem, understanding that problems do not occur in a vacuum; that is, to intellectualize the problem. In the third and fourth phases, the person develops various guiding conjectures or hypotheses that might lead to solutions and reasons through (i.e., critically evaluates) each hypothesis, respectively. The fifth, and final, phase is the testing of fully developed hypotheses, applying those hypotheses that, after critical examination, have been determined to be plausibly valid to the context of the problem.

The fifth phase does not always provide validation, however. In other words, Dewey (1989) acknowledged that the chosen hypothesis might produce failure. But within the habit of reflective activity, failure is not mere failure, but is instructive: "The person who really thinks [reflectively] learns quite as much from [her or] his failures as from [her or] his successes" (p. 206). While outlining "the indispensible traits of reflective thinking" (p. 207), Dewey did not intend to create a fixed or set order. That is to say, the phases could occur in any order, and, depending on the context of the problem, a person could loop through two or three phases repeatedly or extend a particular phase before moving on to another. It is also important to note that Dewey suggested a possible sixth phase: reflectively looking into the future by referencing past experiences.

In an effort to make Dewey's concept of reflective thinking accessible, Rodgers (2002) summarized Dewey's reflective thought into four criteria:

1. Reflection is a meaning-making process that moves a learner from one experience into the next with deeper relationship with and connections to other experiences and ideas. It is the thread that makes continuity of learning possible, and ensures the progress of the individual and, ultimately, society. It is a means to essential moral ends.
2. Reflection is a systematic, rigorous, disciplined way of thinking, with its roots in scientific inquiry.
3. Reflection needs to happen in community, in interaction with others.
4. Reflection requires attitudes that value the personal and intellectual growth of oneself and of others. (p. 845)

Rodgers noted that her accessible criteria were derived primarily from three of Dewey's seminal works: "How We Think" (1989), *Democracy and Education* (1916), and *Experience and Education* (1998). She believed that her criteria demonstrated that reflection is not an end in itself, but a tool used in the transformation of raw experience into meaning-filled theory—grounded in experience and informed by existing theory—to serve the larger purpose of the moral growth of the individual and society (Rodgers, 2002).

APPLYING IDEAS

Reflective Thinking Reflected in the Postmodern

Through using Foucault's reinscription of the concept *discourse*, Derrida's *deconstruction* of language and cultural practices, and Deleuze and Guattari's characterization of the *rhizome*, we extend Dewey's *reflective thinking* here by exploring the initial (written) reflections of seasoned mathematics teachers who completed a graduate-level mathematics education course titled "The Study of Learning and Instruction in Mathematics within the Postmodern." In this section, we are not reporting the results of an empirical study per se, but rather making the argument that exposing mathematics teachers to postmodern theory can assist in changing the face of mathematics teaching, as well as illustrating that the fragmented uncertainty of postmodern thought is important in the current age of surveillance and discipline (Foucault, 1995) through, in the USA, the No Child Left Behind Act of 2001 (Public Law 107–110). We will, however, use extracted quotations from students' reflective essays (the final written assignment from the course) to support our argument that the initial reflections of these seasoned teachers show that they began to reflect differently after their exposure to postmodern theory. We conclude the chapter with a postmodern troubling of the concept *teacher reflection.*

The Course

Teacher reflection was a primary objective as I (the first author) planned the course "The Study of Learning and Instruction in Mathematics within the Postmodern," a graduate-level mathematics education course taught in the May semester of 2005 at Georgia State University, Atlanta, Georgia, USA (the course was also taught in the May semester of 2006 and the spring semester of 2009). A reading intensive discussion seminar, the course began by engaging students in a brief overview of postmodern theory, reading book chapters by foundational French scholars, such as Gilles Deleuze and Félix Guattari (1987), Jacques Derrida (1978), Michel Foucault (1990), and Jean-François Lyotard (1984). In addition, the students read book chapters and essays by education theorists who position their scholarship within postmodern theory, such as Patti Lather (2000), Elizabeth St. Pierre (2000, 2004), and Robin Usher and Richard Edwards (1994). This overview provided the foundation for students to begin an initial critical analysis of essays contained in Margaret Walshaw's (2004a) edited book, *Mathematics Education within the Postmodern*, essays that deconstruct and trouble the discourses of knowledge, learning, teaching, power, equity, and research, among others, within the context of mathematics education (for a review of this book, see Powell, 2007). The course concluded with an initial investigation of the NCTM's *Principles and Standards for School Mathematics* (2000) within the context of the postmodern.

The specific learning objectives of the course were for students to develop an introductory familiarity with the philosophical (i.e., ontological, epistemological, and ethical) underpinnings of postmodern theory and to explore and (re)position the philosophical and structural foundations of mathematics, mathematics teaching and learning, and research in mathematics education within a postmodern framework. The intended purpose was not to "change" the students' teaching practices per se, but rather to provide the opportunity for them as mathematics education professionals to reflect differently on mathematics, mathematics teaching and learning, and, in turn, their pedagogical practices in light of postmodern theory. In short, the purpose of the course was for students to take the familiar discursive binaries of mathematics education (noted previously) and to undergo a deconstructive process, individually and collectively.

Of the twelve students who took the course, ten were part-time graduate students and full-time mathematics teachers: two elementary, two middle school, five high school, and one college. One part-time student was a mathematics teacher professional developer for the state, and the only full-time student taught computer science as a graduate teaching assistant. Eight students were working toward their doctoral degree and four toward their Master's degree. The years of mathematics teaching experience ranged from 5 to 15 years, with teaching experience in urban, suburban, and rural environments, and in private and public schools. The gender and racial (ethnic) makeup of the students were eight female students and four male students, with nine European Americans, two African Americans, and one Caribbean American.

A daily written assignment for the course was to maintain a reading journal (i.e., annotated bibliography) that included written summaries of each assigned reading, student-selected significant quotations from each reading, and comments regarding the student's struggles with each reading and how it might (or might not) assist in her or his teaching (and research). The final for the course was a reflective academic essay (about eight text pages in length) in which each student was to discuss her or his understanding of mathematics education framed in the postmodern and her or his struggles with and remaining (or new) questions of such a framing.

The course was somewhat student-centered, in that students in turn summarized and presented the seminar readings and facilitated the class discussions (students distributed copies of their summaries each day). In addition, scribe notes were taken during each class discussion (the scribe notes were a rotating responsibility completed by each student, with copies distributed the following day). As the instructor of the course, I aimed to construct a Freirian problem-posing, dialogical space where "the teacher is no longer merely the-one-who-teaches, but one who is [herself or] himself taught in dialogue with the students, who in turn while being taught also teach" (Freire, 2000b, p. 80). Freire (2000a), while explicitly defining dialogue, insisted that it must be a "horizontal relationship between persons...[a] relation of 'empathy' between two 'poles' who are engaged in a joint search" (p. 45). Later, he elaborated on the concept of dialogue as he provided the elements that must be present for dialogue to exist. Without expanding on the details, they are: *love, humility, faith, trust, hope*, and *critical thinking* (Freire, 2000b).

Coupled with this problem-posing, dialogical pedagogy, I sought to create a pedagogical space that suspended closure. That is, as the instructor of the course, I constantly reminded the students that a singular "meaning" or "*Truth*" was not to be found in the unfamiliar readings assigned for the class. But rather, by using the readings as a starting point and engaging in class discussions, we might arrive at a multiplicity of meanings and *truths*, both individually and collectively. Feito (2008) recently identified this type of pedagogical space as one that allows *not-knowing* during class discussions. Building on Bakhtin's *Dialogical Imagination*, Feito claimed that to allow the discourse of not-knowing in class discussions assists in developing genuine collaborations among students (and teacher), permitting students (and teacher) to hold divergent and multiple perspectives concurrently and to defer closure. This coupling was an important characteristic of the seminar's pedagogical space, as the students from the course and I, through framing the discursive practices of mathematics teaching and learning and the discipline itself within the postmodern, attempted to "open up the fictions, fantasies and plays of power inherent in mathematics education" (Walkerdine, 2004, p. viii). In the discussion that follows, we shed light on the impact of this postmodern opening up of mathematics and mathematics teaching and learning as illustrated in the students' reflective academic essays.

Teachers Reflecting *Differently*

Postmodern theory "has crept in and taken hold of my comforting mathematical discourses and deconstructed them from underneath me," wrote Nicholas (pseudonym, as are all proper names throughout), one of the Master's students, in his reflective essay. "While in the beginning I resisted," Nicholas continued, "I have now allowed myself to follow down some of these postmodern paths and seen the positive possibilities." On the other hand, Lauren, a doctoral student, noted, "It was this 'morality of uncomfortableness' that actually leads me to feel comforted by postmodernism.... There are others who recognize the complexities of social relations, who actually have named the condition of life as I see it." Similarly, Brook, another doctoral student, wrote, "In postmodernism I found a language, previously unavailable to me, to describe aspects of my personal beliefs and practice that before seemed a pseudo-random collection of ways of operating in the world. In many ways... postmodernism was a natural fit." The dichotomy of these students' statements illustrates the range of attitudes (or comfort) toward postmodern theory, as each attempted to articulate her or his understanding(s) or, better yet, not-knowing. Within this range of attitudes, however, were students who emphatically questioned the possibility of the postmodern within education altogether. For instance, Deanne, a doctoral student, wrote, "From a realistic standpoint, the educational system will never truly be postmodern; it will always be governed by state and local powers that base success of a school upon tests and how many students from each sub-category take them." Deanne, nonetheless, also began to reflect differently, writing, "Teachers can challenge this discourse [of testing] by the decisions we make in our classes, for our students, every day."

No matter what the students' initial comfort level with the ideas of postmodern theory, in the following discussion, we argue that their final reflective essays demonstrate that, in most cases, each student's thinking attempted to take a new "line of flight," in which they endeavored to "make a map and not a tracing" (Deleuze & Guattari, 1987, pp. 11–12) of the meanings and truths of mathematics teaching and learning. Through using the first phase of Dewey's (1989) five phases of reflecting thinking—suggestion—

the discussion attempts to capture (some of) these new lines of flight, illustrating how these reflective practitioners (Schön, 1983) at times begin to reflectively think differently (or not). We focus the discussion on Dewey's first phase because the other four phases—intellectualization, guiding hypothesis, reasoning, and testing the hypothesis by action—would require further, long-term contact with the students to document how they contextualized the different suggestions, how their different reflective thinking transformed (or not) the developing of and reasoning through different hypotheses, and how various hypotheses were put into action.

In other words, a caveat is necessary as we begin the discussion, in that the discussion is not about tracking or documenting mathematics "teacher change." We understand mathematics teacher change to be a complex endeavor (see Brown & Borko, 1992; Fennema & Nelson, 1997). Like the NCTM documents, we believe that teaching is a continual journey, in that "effective teachers" do not master teaching, but rather find themselves in a continuous state of growth and change (Mewborn, 2003). Or, said in another way, effective teachers find themselves in a continuous state of becoming. Consequently, becoming a teacher, as described by Gomez, Black, and Allen (2007), is a process that is never finalized or fixed, but rather a fluid process of continuous critical examination of self and students in which old ways of thinking and acting are disrupted and transformed into new ways of thinking and acting. It is this continuous state of becoming that the following brief discussion explores.

Within Dewey's (1989) reflective thinking phase of "suggestions, in which the mind leaps forward to a possible solution" (p. 200), we believe that postmodern theory offered these seasoned teachers the possibility of different suggestions, as the familiar discursive binaries of mathematics education underwent deconstruction, and, in turn, motivated different suggestions. Mila, a Master's student, wrote, "Postmodernism is not trying to lead one to believe that the ideals encompassed within it are absolute; it is mere suggestions of alternative ways of thinking and pursuance, so that certain discourses can be negotiated." In this instance, Mila reflects differently on the discursive practices of mathematics education, acknowledging that the discursive practices can, indeed, be negotiated and, as Hardy (2004) suggested, that the "modifier 'discursive' stresses the ways in which all practices are bound up in systems of knowledge" (p. 106). Sarah, a Master's student, noted, "As teachers, I feel we must reflect on our own discourses . . . and work through and past them in the quest for something better for our students." Hardy (2004) claimed that acknowledging mathematics education classifications and descriptions as *discursive practices* enables us to become aware of how these classifications and descriptions might be described differently.

For instance, in a postmodern frame, new suggestions emerge that begin to classify differently or pry loose the discursive binary teacher/student, deconstructing it, both in teacher and student identity and relations of power. Within postmodern theory, teachers and students are reinscribed as *subjects*, rather than as *individuals*. The term *individual* is a humanist term that implies that there is an "independent and rational being who is predisposed to be motivated toward social agency and emancipation—what Descartes believed to be the existence of a unified self" (Leistyna, Woodrum, & Sherblom, 1996, p. 341). Postmodern theory rejects this humanist notion of an essential, unified self who is always present, because it minimizes the force of social structures and discourses on the person. This humanist perspective of the person virtually denies that structures and discourses have any impact at all on the formation of the individual. Rather, a postmodern perspective reinscribes the person as a multiplicitous and fragmented subject who is subjugated, but not determined, by the structures and discourses that constitute the person, mapping out a "territory...in which structure and agency are not either-or but both-and and, simultaneously, neithernor" (Lather, 1991, p. 154).

This postmodern reinscription of teachers and students as subjects provides for a different suggestion of power and, in turn, agency. *Power*, in a postmodern frame, is reinscribed as a dynamic and productive event that exists in *relations* (Foucault, 1990), not as an object that can be acquired, seized, shared, or that can slip away. Therefore, rather than speaking of power, Foucault spoke of "relations of power" or "power relations" (p. 94). For Foucault, power relations are a multiplicity of force relations that constitute their own organization; a process of struggles and confrontations that transforms, strengthens, or reverses the relations; the points of support or resistance of a system; and/or the strategies that design and maintain social structures and discourses. Foucault further claimed that "where there is power, there is resistance"; indeed, "there is a plurality of resistances, each of them a special case" (pp. 95–96). That is to say, in this Foucauldian reinscription of power, resistance can be achieved not only by the united actions of working men from all countries (Marx & Engels, 1978), but also by the solitary actions of the discursively constituted subject.

Deanne used this Foucauldian reinscription of power and agency when she argued that teachers can challenge discourses by the decisions they make in their classes, for their students, every day (as noted previously). Deanne expanded her argument, writing, "Teachers in a postmodern classroom (occupied by subjects who transfer power between the teacher and each other in order to gain knowledge) attempt to create a space where students [and teachers] can learn through communication with others in the class." Similarly, Lauren wrote, "I must consciously acknowledge my students, not as objects, but as [subjects], using power, resisting power, and interacting with each other and with the mathematics." While reconstituting power as "'letting go' of the control in their classroom" and allowing "for the possibility of being 'found out' as not being *the* authority," Charles, a doctoral student, wrote, "Teachers need to embrace their lack of expertise. . . . By joining the learning process in the classroom, teachers can model the open-mindedness necessary for students so that they might begin questioning, discussing, and constructing their own mathematical knowledge." This joining in the learning process allows for a different interaction between teacher and students—and mathematics—that supports the mathematics classroom in becoming a pedagogical space that is open for "*negotiation of intentionality*" (Valero, 2004, p. 49, emphasis in original).

Valero (2004) suggested that when students (and teachers) are defined as agents who negotiate the intentions of the mathematics classroom—using power, resisting power, interacting with each other and the mathematics—that real empowerment might take place. Here, empowerment is understood as *self*-empowerment: "a process one undertakes for oneself; it is not something done 'to' or 'for' someone" (Lather, 1991, p. 4). Within the context of a postmodern mathematics classroom, Valero claimed that empowerment is not passed from teacher to student through the transference of "powerful knowledge," but rather might be defined in terms of the potentialities for students (and teachers) to participate in (i.e., to negotiate) the discursive practices of school mathematics. Sarah noted, "I hope to help my students empower themselves to overcome the discourses . . . to overcome the limitations society and our culture has put on them."

Coupled with this different understanding of student and teacher empowerment was a different suggestion of understanding students and teachers as multiplicitous and fragmented subjects. Within the suggestion of multiplicitous fragmentation, Lauren wrote, "I have been many in my life—there is no one *woman* who defines me. I am mother, wife, teacher, daughter, boss, and student—each time made anew by social context and relationships with others." As these seasoned teachers began to understand themselves as multiplicitous and fragmented subjects, constructed through discourses and relations of power, they, in turn, began to view their students as multiplicitous and fragmented. For example, Nancy, a doctoral student, stated: "Educators should begin to look at their students as multiplicitous subjects, rather than as individuals; it is important to remember students are not identical in math or English class, in sports or hobbies, at home or school." She continued, writing, "I need to accept my students as multiplicitous—each one coming to me with different levels of prior mathematical knowledge and different ways of learning." Susan, a doctoral student, wrote, "If nothing else, I have come out of this class knowing that students think differently, react differently, and position themselves differently; I need to recognize and respect these multiplicities."

Fleener (2004), building on Deleuze and Guattari's (1987) rhizome, argued for the importance of seeing teachers and students (and the mathematics curriculum) as multiplicitous: "Viewing the role of teacher as multiplicitous supports identity multiplicity. Understanding students as multiplicitous is important when we try to make sense of their differences in understandings, interests, and performances" (p. 210). She further suggested that teachers who see themselves, students, and curriculum as multiplicitous shift their "focus to the in-between, the relational, and the dynamic" (p. 213). Through "engaging the in-between, students build their own understanding, not as foundations, but as complex webs of the nexus of relationship in the abstract world of mathematics" (p. 214). Teachers in-between see their interactions with students in the mathematics classroom with "soft eyes," looking beyond standards, end-of-course tests, or daily learning objectives, as they enact *with* students the experience of mathematical exploration and creation (Fleener, 2004). Sarah began looking with soft eyes, writing, "Typically, in mathematics we think there is one right answer to a problem and focus on developing our students' knowledge of how to get to that answer. In postmodern thought, it is not just about the problem or the answer, but the 'in-between' that matters the most."

A new suggestion of the in-between brought about a different suggestion regarding the possibilities of classroom communication. Within the postmodern, Cabral (2004) claimed, language ceases to be regarded as a means of "communication," but as the very process of constitution of the subject; that is, the discipline of mathematics, students, and teachers are constituted within a language community. Therefore, Cabral argued, "we need to stop talking and start listening to the student . . . it is through speaking that one learns and through listening that one teaches" (p. 147). Lauren wrote, "I will listen more, talk less. . . . Let the students guide the lesson, hear what they have to say, to me and to each other, about the mathematics, about their understandings, questions, and confusions." Nancy noted, "Actively listening to students' questions and concerns may lead to further areas of exploration outside of the daily . . . lesson." Dorothy, a doctoral student, spoke about the importance of teachers listening to their students, and of students listening to each other: "There have been many times in my classroom when I could not understand the point a student was trying to make. It took another student, in different words, to relay the message so that I could understand." Macmillan (2004) suggested that teachers who develop classrooms where student-talk is valued equally with teacher-talk allow students to bridge the gap between where students are in their understandings and perceptions, and where the teacher is. In short, such classrooms allow "knowledgeable children *to be*, rather than having *to be taught*" (p. 95, emphasis in original).

The preceding brief discussion attempted to capture the different suggestions that engaging in the postmodern provided these seasoned teachers as they began to think differently about the discursive binary of teacher/student. These suggestions motivated different classifications and descriptions for teachers' and students' identity, agency, and empowerment, and, in turn, a different suggestion of teacher and student participation in the mathematics classroom. There were several other instances in the teachers' final essays in which other familiar discursive binaries were deconstructed or troubled. Some troubled the binary of mathematical-able student/non-mathematical-able student, while others troubled the effective teacher/non-effective teacher binary. And, in rare occasions, even the discursive binary of mathematical Truths/mathematical truths was troubled. For instance, Marcus, a doctoral student, wrote, "Are we confining ourselves and our students by the rules and laws of mathematics that do not allow for them to do the unexpected, to go beyond their own reality?" Likewise, Nicholas noted, "I was blown away by the thought that mathematics, something that I had found comfort in because of its absolute nature, was being viewed as a science of uncertainty that could not be defined by its absolutes any longer." In general, the teachers limited their comments regarding the truths of mathematics, or, similar to Nicholas, somehow resisted reconstituting the "absolute nature" of mathematics. It appears that although mathematics has been argued to be the roots of postmodern thought (see Tasić, 2001), to deconstruct the capital-T truths of mathematics might prove to be the most difficult deconstruction to undertake; it may be, nonetheless, the most important.

Furthermore, even though the discussion focused on only one phase of Dewey's (1989) reflective thinking (suggestion) and only one discursive binary (teacher/student), we believe that the discussion nonetheless demonstrated that the teachers' initial reflections began to reflect differently in light of postmodern theory, and were somewhat consistent with and extended the four criteria of Dewey's reflective thought, as outlined by Rodgers (2002). First, the teachers' comments illustrated that they did move from one reconstituted experience to the next, making deeper relationships with and connections to other experiences. Second, their comments were rigorous; that is to say, they consistently connected their reconstituted experiences to the readings from the course. Not only did they connect their multiple and fragmented meanings of the different suggestions to the readings, but also to the class discussions, illustrating, third, that reflection does, indeed, happen in interaction with others. And last, through their final essays, they demonstrated different personal and professional attitudes that valued not only their own personal and intellectual growth, but also the personal and intellectual growth of their peers and their students. Sarah's concluding comments from her essay captures Rodger's four criteria of reflective thought in the postmodern succinctly:

Postmodernists do not desire to just tear down the structures that are in place, but to determine what had a place in setting up the structures that are there and work at rebuilding a more just society. Before we can break the cycle of what is going on in education: academic tracking; racial, economic, and gender inequality; and teacher-centered instruction, to name a few, we have to understand what caused those things to exist to begin with. With that understanding, we might be able to deconstruct those things and build an educational system that promotes multiple ways of teaching and learning, and one that ensures that all students receive an equitable education. As a mathematics educator, I can begin to do such deconstruction in my own classroom and share my experiences, insight, and hopes with other teachers. It is by doing this that I might aid in that step toward helping education make the turn that is needed to facilitate all learning and actually leave no child behind.

Deconstructing the Discourse of Teacher Reflection

Throughout this chapter we have privileged the discourse "teacher reflection." To keep consistent with a postmodern skepticism, we conclude the chapter with a brief deconstruction or troubling of the concept, in hopes of drawing "attention to our ultimately compromised stance in everything we do and say" (Walshaw, 2004b, p. 11). At the risk of being circular, we must reflect on—and trouble—how we use reflection (E. A. St. Pierre, personal communication, August 2008).

Given that the concept *reflective thinking* has become sloganized, it might be obscuring more than it reveals about its values and stances (Noffke & Brennan, 1988). In other words, as Fendler (2003) claimed, "Cartesian rationality, Deweyan educational aims, Schönian professionalism, and individual agency endow reflective thinking with a seductive appeal that has tended to deflect critical appraisal" (p. 22). From a postmodern stance, the very idea of a Cartesian self that exists outside of and/or separate from a socially constructed one, and therefore can stand apart and reflect on itself, must be called into question. For instance, "How can [we] assume that society is structured by forces of domination and oppression and at the same time promote reflective thinking as if it had not also been shaped by those forces of oppression?" (Fendler, 2003, p. 20). We might be inadvertently furthering that oppression through our reification of reflective thinking. Fendler pointed out, "It is ironic that the rhetoric about reflective practitioners focuses on empowering teachers, but the requirements of learning to be reflective are based on the assumption that teachers are incapable of reflection without direction from expert authorities" (p. 23). In some sense, we may be said to be using the master's tools to unmake his house, seeing no better alternative (Delpit, 1995).

Having accepted the possibility of reflection, we must be wary of how it is used. We cannot see reflection as an end in and of itself, but rather as a means to develop more ethical judgments and strategic action towards ethically important ends (Noffke & Brennan, 1988). But who decides what those ends are? We must acknowledge that assigning specific readings and privileging postmodern understandings might be "prompting student reflections using a discourse of difference that, in fact, prevents the use of difference" (J. Fleener, personal communication, April 2009). The reflective thoughts of students are, themselves, discursive formations influenced by the goals of the course. The situation, however, is surely much more complex and nondeterministic than that: "Education can never simply be understood as a process where the teacher moulds the student" (Biesta, 2005, p. 6). While I, the first author and the instructor of the course, did have an agenda, I also attempted to put that agenda under scrutiny by allowing not-knowing. Not to expose students to postmodernism is to perpetuate the status quo; to expose them to postmodernism is not the same as forcing them to accept it. All I have done, I hope, is given the students one more choice than they had before. It is up to them to take different lines of flight, or not.

Moreover, the connotation of *reflection* as merely looking at one's self in a mirror can lead to the inadvertent omission of the action component necessary for reflection to have any effect. As we did not pursue the students from the course through their reflective (postmodern) journey, we cannot say with confidence that their newfound language or postmodern attitude will actually affect their teaching. I, as the instructor of the course, however, can hope that by engaging mathematics teachers in the postmodern, I have begun to disrupt (or trouble) some of the discourses that they are often unconsciously burdened with, and helped them to think differently—as I (we), in turn, trouble my (our) own efforts. In short, our collective perspective regarding our compromised stance in everything we do and say is similar to Foucault's (1997), who said: "My point is not that everything is bad, but that everything is dangerous, which is not exactly the same as bad. If everything is dangerous, then we always have something to do. So my position leads not to apathy but to hyper- and pessimistic activism" (p. 256).

REFERENCES

Biesta, G. J. J. (2005). What can critical pedagogy learn from postmodernism? Further reflections on the impossible future of critical pedagogy. In Ilan Gur Ze'eve (Ed.), *Critical theory and critical pedagogy today: Toward a new critical language in education* (pp. 143–159). Haifa, IL: University of Haifa Press.

Brown, C. A., & Borko, H. (1992). Becoming a mathematics teacher. In D. A. Grouws (Ed.), *Handbook of research on mathematics teaching and learning* (pp. 209–239). New York: Macmillan.

Cabral, T. (2004). Affect and cognition in a pedagogical transference: A Lacanian perspective. In M. Walshaw (Ed.), *Mathematics education within the postmodern* (pp. 141–158). Greenwich, CT: Information Age Press.

Davis, P. J., & Hersh, R. (1981). *The mathematical experience* (First Mariner Books ed.). Boston: Houghton Mifflin Company.

Deleuze, G., & Guattari, F. (1987). Introduction: Rhizome (Trans: B. Massumi). *A thousand plateaus: Capitalism and schizophrenia* (pp. 3–25). Minneapolis, MN: University of Minnesota Press. (Original work published 1980)

Delpit, L. (1995). *Other people's children: Cultural conflict in the classroom.* New York: The New Press.

Derrida, J. (1978). Structure, sign and play in the discourse of the human sciences (Trans: A. Bass). *Writing and difference* (pp. 278–293). Chicago: University of Chicago Press. (Lecture delivered 1966)

Derrida, J. (1997). *Of grammatology* (Trans: G. C. Spivak, Corrected ed.). Baltimore, MD: Johns Hopkins University Press. (Original work published 1974)

Derrida, J., & Montefiore, A. (1992). *Jacques Derrida: Wall to Wall Television Production for Channel 4* [videorecording]. Princeton, NJ: Films for the Humanities.

Derrida, J., & Montefiore, A. (2001). "Talking liberties": Jacques Derrida's interview with Alan Montefiore. In G. Biesta & D. Egéa-Kuehne (Eds.), *Derrida & education* (pp. 176–185). New York: Routledge.

Dewey, J. (1916). *Democracy and education: An introduction to the philosophy of education.* New York: The Free Press.

Dewey, J. (1974). The relation to theory to practice in education. In R. D. Archambault (Ed.), *John Dewey on education: Selected writings* (University of Chicago ed., pp. 313–338). Chicago: University of Chicago Press. (Original work published 1904)

Dewey, J. (1989). How we think. In J. A. Boydston (Ed.), *John Dewey: The later works, 1925–1953* (Vol. 8, pp. 105–352). Carbondale, IL: Southern Illinois University Press. (Original work published 1933)

Dewey, J. (1998). *Experience and education: The 60th anniversary edition* (60th anniversary ed.). West Lafayette, IN: Kappa Delta Pi. (Original work published 1938)

Dillon, M. C., & Derrida, J. (1999). In R. Audi (Ed.), *The Cambridge dictionary of philosophy* (2nd ed., p. 223). Cambridge, UK: Cambridge University Press.

Ernest, P. (1998). *Social constructivism as a philosophy of mathematics.* Albany, NY: State University of New York Press.

Feito, J., A. (2008). Allowing not-knowing in a dialogic discussion. *International Journal for the Scholarship of Teaching and Learning.* Retrieved August 19, 2008, from http://academics.georgiasouthern.edu/ijsotl/v1n1/feito/ij_feito.htm

Fendler, L. (2003). Teacher reflection in a hall of mirrors: Historical influences and political reverberations. *Educational Researcher, 32*(3), 16–25.

Fennema, E., & Nelson, B. S. (Eds.). (1997). *Mathematics teachers in transition.* Mahwah, NJ: Lawrence Erlbaum Associates.

Fleener, M. J. (2004). Why mathematics? Insights from poststructural topologies. In M. Walshaw (Ed.), *Mathematics education within the postmodern* (pp. 201–218). Greenwich, CT: Information Age Press.

Foucault, M. (1972). *The archaeology of knowledge* (Trans: A. M. Sheridan Smith.). New York: Pantheon Books. (Original work published 1969)

Foucault, M. (1990). *The history of sexuality. Volume I: An introduction* (Trans: R. Hurley). New York: Vintage Books. (Original work published 1976)

Foucault, M. (1995). *Discipline and punish: The birth of the prison* (Trans: A. Sheridan). New York: Vintage Books. (Original work published 1975)

Foucault, M. (1997). On the genealogy of ethics: An overview of work in progress. In P. Rabinow (Ed.), *The essential works of Michel Foucault, 1954–1984* (Vol. I, Ethics, pp. 253–280). New York: New Press. (Interview conducted 1983)

Freire, P. (2000a). *Education for critical consciousness* (Continuum ed.). New York: Continuum. (Original work published 1969)

Freire, P. (2000b). *Pedagogy of the oppressed* (Trans: M. B. Ramos, 30th anniversary ed.). New York: Continuum. (Original work published 1970)

Gomez, M. L., Black, R. W., & Allen, A.-R. (2007). "Becoming" a teacher. *Teachers College Record, 109*(9), 2107–2135.

Gordon, C. (1980). Preface (Trans: C. Gordon, L. Marshall, J. Mepham, & K. Soper). In C. Gordon (Ed.), *Power/knowledge: Selected interviews and other writings, 1972–1977 by Michel Foucault* (pp. vii–x). New York: Pantheon Books.

Hardy, T. (2004). "There's no hiding place": Foucault's notion of normalization at work in a mathematics lesson. In M. Walshaw (Ed.), *Mathematics education within the postmodern* (pp. 103–119). Greenwich, CT: Information Age Press.

Hart, L. C., Najee-ullah, D., & Schultz, K. (2004). The reflective teaching model: A professional development model for in-service mathematics teachers. In R. N. P. Rubenstein & G. W. Bright (Eds.), *Perspectives on the teaching of mathematics* (pp. 207–218). Reston, VA: National Council of Teachers of Mathematics.

Hiebert, J. (2003). What research says about the NCTM Standards. In J. Kilpatrick, W. G. Martin, & D. Schifter (Eds.), *A research companion to Principles and Standards for School Mathematics* (pp. 5–23). Reston, VA: National Council of Teachers of Mathematics.

Jaworski, B. (1998). Mathematics teacher research: Process, practice and the development of teaching. *Journal of Mathematics Teacher Education, 1*, 3–31.

Kilpatrick, J., Martin, W. G., & Schifter, D. (Eds.). (2003). *A research companion to principles and standards for school mathematics*. Reston, VA: National Council of Teachers of Mathematics.

Lather, P. (1991). *Getting smart: Feminist research and pedagogy with/in the postmodern*. New York: Routledge.

Lather, P. (2000). Drawing the line at angels: Working the ruins of feminist ethnography. In E. A. St. Pierre & W. S. Pillow (Eds.), *Working the ruins: Feminist poststructural theory and methods in education* (pp. 284–311). New York: Routledge.

Leistyna, P., Woodrum, A., & Sherblom, S. A. (Eds.). (1996). *Breaking free: The transformative power of critical pedagogy*. Cambridge, MA: Harvard Educational Review.

Lyotard, J. F. (1984). *The postmodern condition: A report on knowledge* (Trans: G. Bennington & B. Massumi). Minneapolis, MN: University of Minnesota Press. (Original work published 1979)

Macmillan, A. (2004). Facilitating access and agency within the discourses and culture of beginning school. In M. Walshaw (Ed.), *Mathematics education within the postmodern* (pp. 77–101). Greenwich, CT: Information Age Press.

Marx, K., & Engels, F. (1978). Manifesto of the Communist Party. In R. C. Tucker (Ed.), *The Marx-Engels reader* (2nd ed., pp. 469–500). New York: Norton. (Original work published 1848)

Mewborn, D. S. (1999). Reflective thinking among preservice elementary mathematics teachers. *Journal for Research in Mathematics Education, 30*(3), 316–341.

Mewborn, D. S. (2003). Teaching, teachers' knowledge, and their professional development. In J. Kilpatrick, G. Martin, & D. Schifter (Eds.), *A research companion for NCTM Standards* (pp. 45–52). Reston, VA: National Council for Teachers of Mathematics.

National Council of Teachers of Mathematics. (1989). *Curriculum and evaluation standards for school mathematics.* Reston, VA: National Council of Teachers of Mathematics.

National Council of Teachers of Mathematics. (1991). *Professional standards for teaching mathematics.* Reston, VA: National Council of Teachers of Mathematics.

National Council of Teachers of Mathematics. (1995). *Assessment standards for school mathematics.* Reston, VA: National Council of Teachers of Mathematics.

National Council of Teachers of Mathematics. (2000). *Principles and standards for school mathematics.* Reston, VA: National Council of Teachers of Mathematics.

Noffke, S. E., & Brennan, M. (1988). *The dimensions of reflection: A conceptual and contextual analysis.* New Orleans, LA: American Educational Research Association. (ERIC Document Reproduction Service No. ED296969)

Powell, G. C. (2007). The view from here: Opening up postmodern vistas [Review of the book *Mathematics education within the postmodern*]. *The Mathematics Educator, 17*(1), 42–44.

Rodgers, C. (2002). Defining reflection: Another look at John Dewey and reflective thinking. *Teachers College Record, 104*(4), 842–866.

Sarup, M. (1993). *An introductory guide to post-structuralism and postmodernism* (2nd ed.). Athens, GA: University of Georgia Press.

Schoenfeld, A. (2004). The math wars. *Educational Policy, 18*(1), 253–286.

Schön, D. A. (1983). *The reflective practitioner: How professionals think in action.* New York: Basic Books.

Seidman, S. (1994). Introduction. In S. Seidman (Ed.), *The postmodern turn: New perspectives on social theory* (pp. 1–23). Cambridge, UK: Cambridge University Press.

St. Pierre, E. A. (2000). Poststructural feminism in education: An overview. *International Journal of Qualitative Studies in Education, 13*(5), 467–515.

St. Pierre, E. A. (2004). Care of the self: The subject and freedom. In B. Baker & K. E. Heyning (Eds.), *Dangerous coagulations?: The uses of Foucault in the study of education* (pp. 325–358). New York: Peter Lang.

Tasić, V. (2001). *Mathematics and the roots of postmodern thought.* Oxford, UK: Oxford University Press.

Thomas, C. D., & Santiago, C. A. (2006). Meaningful mathematics for urban learners: Perspective from a teacher-researcher collaboration. In J. D. Masingila

(Ed.), *Teachers engaged in research: Inquiry into mathematics classrooms, grades 6–8* (pp. 147–169). Reston: VA: National Council of Teachers of Mathematics.

Usher, R., & Edwards, R. (1994). *Postmodernism and education.* NY: Routledge.

Valero, P. (2004). Postmodernism as an attitude of critique to dominant mathematics education research. In M. Walshaw (Ed.), *Mathematics education within the postmodern* (pp. 35–54). Greenwich, CT: Information Age Press.

Walkerdine, V. (2004). Preface. In M. Walshaw (Ed.), *Mathematics education within the postmodern* (pp. vii–viii). Greenwich, CT: Information Age Press.

Walshaw, M. (Ed.). (2004a). *Mathematics education within the postmodern.* Greenwich, CT: Information Age Press.

Walshaw, M. (2004b). Introduction: Postmodern meets mathematics education. In M. Walshaw (Ed.), *Mathematics education within the postmodern* (pp. 1–11). Greenwich, CT: Information Age Press.

Walshaw, M. (2007). *Working with Foucault in education.* Rotterdam, NL: Sense.

Wilson, P. S., Cooney, T. J., & Stinson, D. W. (2005). What constitutes good mathematics teaching and how it develops: Nine high school teachers' perspectives. *Journal for Mathematics Teacher Education, 8,* 83–111.

Wilson, M., & Goldenberg, M. (1998). Some conceptions are difficult to change: One middle school mathematics teacher's struggle. *Journal of Mathematics Teacher Education, 1,* 269–293.

FURTHER READING

Dreyfus, H. L., Rabinow, P., & Foucault, M. (1983). *Michel Foucault, beyond structuralism and hermeneutics* (2nd ed.). Chicago: University of Chicago Press.

Peters, M., & Burbules, N. C. (2004). *Poststructuralism and educational research.* Lanham, MD: Rowman & Littlefield.

Seidman, S. (Ed.). (1994). *The postmodern turn: New perspectives on social theory.* Cambridge, UK: Cambridge University Press.

Usher, R., & Edwards, R. (1994). *Postmodernism and education.* London: Routledge.

Walshaw, M. (Ed.). (2004). *Mathematics education within the postmodern.* Greenwich, CT: Information Age Press.

Walshaw, M. (2007). *Working with Foucault in education.* Rotterdam: Sense Publishers.

CHAPTER 11

LEARNING THROUGH DIGITAL TECHNOLOGIES

Nigel Calder and Tony Brown

ABSTRACT

How do pedagogical media transform the mathematical phenomena to which they seek to grant access? This chapter examines the ways in which understanding emerges when mathematical phenomena are engaged through digital pedagogical media—the spreadsheet, in particular. We adopt a hermeneutic perspective on both learning and research processes, within a depiction of an evolving mathematics education research landscape. We then consider the alternative insights into teaching and learning processes such a hermeneutic frame presents. We apply these ideas to data drawn from a study involving 10-year-old students, to illustrate how their learning was fashioned by the particular affordances of the pedagogical media. Further pedagogical implications associated with this influence will also be offered.

INTRODUCING IDEAS

How do we understand the mathematical learning process? The preconceptions that each learner brings to mathematical phenomena, and activity associated with it, are derived from the specific cultural domain that the

Unpacking Pedagogy: New Perspectives for Mathematics Classrooms, pages 223–243

learner inhabits. Within a hermeneutic perspective, mathematical learning comprises a process of interpretation, where understanding and "concepts" might be seen as states caught in ongoing formation, rather than as once-and-for-all fixed realities. Our understandings of the mathematical phenomena, and of who we are, evolve through cyclical engagements with the phenomena and the constant drawing forward of prior experiences and understandings. Here, "mathematical concepts" are not fixed realities, but rather more elusive, formative processes that become further enriched as the learner views events from fresh, ever-evolving perspectives. The pedagogical medium, the mathematical task, the pre-conceptions of the learners, and the associated dialogue evoked are interdependent and co-formative. It is from their relationship with the learner that understanding emerges (Calder, Brown, Hanley, & Darby, 2006). This understanding is their interpretation of the situation through those various filters. Understanding emerges from cycles of interpretation, but this is forever in transition: There may always be another interpretation made from the modified stance (Brown, 2001; Calder, 2007).

Central to this interpretive process is the hermeneutic circle, which combines notions of language and structure in emphasizing interpretation through the development of individual explanations (Gadamer, 1989). The learner develops explanations based on his or her interpretations of the phenomena. These explanations then encounter resistance from broader discourses, so that understanding evolves and the explanations shift. Here, there is always a gap between the interpretation and the explanation, and this provides the space for understanding and learning to occur. The circularity between present understanding and explanation (where the explanation gives rise to a change in perspective) in turn evokes a new understanding. The interpreter's attention moves cyclically from the part, to the whole, to the part, and so forth, until some manner of resolution or consensus emerges. Within the learning context, the whole can be aligned with the various discourses or schema the learner brings to the situation, and the part can be aligned with the specificity of the situation they confront (perhaps in the form of a learning activity). The engagement of learners oscillates between their prevailing discourse and the activity. With each of these iterations, their perspective alters, and as they re-engage with the activity from these fresh perspectives, their understanding evolves. Ricoeur's (1981) notion of the hermeneutic circle, which guides us here, emphasizes the interplay between understanding and the narrative framework within which this understanding is expressed discursively, and which helps to fix it. While these "fixes" are temporary; they orientate the understanding that follows and the way this comes to be expressed. In seeing understanding as linguistically based, the student dialogue and comment provides the source for the interpretations of their mathematical understanding.

In the research that follows, the evolving history of the learner is a collaboration of their dialogue and the corresponding action. A hermeneutic viewpoint allows the incorporation of dialogue and actions, as the links between what was being said or written, and the participants' investigative approach, were examined in terms of their interpretation of the mathematical phenomena. As well, the data were hinged to the discourse that constituted their production and analyses. This perspective "begins with the problem of unmediated access to a *transparent* mathematical reality, shifting the emphasis from the critical learner as the site of original presence, to a decentred relational complex process" (Walshaw, 2001, p. 28). Consequently, by varying the pedagogical medium, alternative frames are generated, hence rendering the learning experiences and ensuing dialogue in a different manner and allowing space for the restructuring of mathematical understanding—for alternative ways of knowing (Brown, 2008a, 2008b). Thinking such as this challenges the notion of constructed, abstract concepts being transposed intact across varying contexts.

In this chapter, we are specifically concerned with the learner's preconceptions of pedagogical media, and how these, in conjunction with the opportunities and constraints offered by the media itself, promote distinct pathways in the learning process. That is, mathematical activity is inseparable from the pedagogical device, as it were, derived as it is from a particular understanding of social organization. Hence, the mathematical ideas developed will inevitably be influenced by this device. This pedagogical device is more than an environment. It is imbued with a complexity of relationships evoked by the users and the influence of underlying discourses. For example, in some Vygotskian accounts of mathematical learning, tools, such as linguistic constructs, are seen as acting as mediators situating the learning with reference to particular traditions (e.g., Lerman, 2006). Research involving the utilization of Information and Communication Technologies (ICT) in mathematics education often utilize this frame in accounting for alternative cognitive internalization through the mediation of cultural tools (e.g., Arzarello, Paola & Robutti, 2006; Marriotti, 2006). Confrey and Kazak (2006), likewise, have argued that learning in mathematics involves both activity and socio-cultural communication interacting in significant ways. They contend that neither influence is privileged, nor, in fact, can be separated, as we are simultaneously participants and observers in all enterprises, at all times. This view resonates with the hermeneutic perspective that framed this research, in which the participants' actions and interpretations were taken as interacting filters in the understanding that emerged. In a similar manner, the objectification of understanding can be perceived as being underpinned by the interplay of *typological* meaning (language) and *topological* meaning (visual figures and motor gestures) (Radford, Bardini, & Sabena, 2007).

In the study to be outlined in the following section, an analysis was undertaken into the ways participants' preconceptions in mathematical thinking were reorganized by engagement through the pedagogical medium of the spreadsheet. The manner in which these new perspectives then framed any subsequent re-engagements, and the way in which participants' learning trajectories were influenced by the pedagogical medium, were both considered. Likewise, the dialogue evoked by the engagement was examined to ascertain ways it may have led to alternative conceptualization and understanding. The following section interrogates the learning process using research data viewed through the hermeneutic lens. If the pedagogical medium elicited alternative ways of knowing, we also need to consider the implications of using digital technologies on the social organization of the classroom.

APPLYING IDEAS

The research described is part of an ongoing research program involving primary-aged pupils and pre-service teacher students, exploring how spreadsheets might function as pedagogical media. The participants in these excerpts were drawn from 10-year-old students, attending five schools associated with a university in New Zealand. They were, at the time, involved in a collaborative project offering programs to develop gifted and talented students in their schools. This particular group comprised twenty-one students (twelve boys and nine girls) who had been identified through a combination of problem-solving assessments and their teachers' recommendations. The students came from a range of socio-economic backgrounds. They were brought together and situated in a classroom that included seven computers with spreadsheets as available software. This was the typical working environment for two of the schools, while the other three schools had three or four computers in each class at this level. For the students from those three schools, the computer access during the research was, therefore, marginally enhanced, compared with that available in their usual classroom situation.

For the research project, the students worked on a program of activities, using spreadsheets to investigate mathematical problems predominantly deemed suitable for developing algebraic thinking. They were observed, their conversations were recorded and transcribed, and their investigations were printed out or recorded. There were school group interviews, and interviews with working pairs. They undertook a survey based on opinion and motivation. Ongoing data were also gathered over a longer-term period (eighteen months) with three of the groupings, and this allowed for some case study data to emerge. Observations, and the recording and transcribing of participants' conversations in subsequent groups from the schools, also further enriched the data set.

The research questions for this study centered upon the ways that the pedagogical medium of the spreadsheet influenced participants' learning experiences. In particular, the following were examined: In what ways might the learning experiences differ when students use spreadsheets to investigate mathematical problems? How might investigating mathematical phenomena through digital pedagogical media produce alternative conceptualizations of the mathematics involved? Allied to this, of interest were the understandings that emerged for the students through the specific learning environment. The inquiry attended to understandings and meanings, and with context profoundly implicated in meaning, a natural setting was set up. However, the intrusion and associated influence of the researcher was inevitable (see Mason, 2002). As such, the researcher became a constituent of the data. An aim was to minimize the intrusion, and while this presence would exert some influence, any ensuing effect was not the focus of the observations.

Participants were involved in the following procedures:

- Observations
- Activities using spreadsheets, as part of their program
- Individual assessment tasks
- Interviews
- Questionnaires

Illustration of the Hermeneutic Circle

The following excerpt from the data illustrates how a hermeneutic circle models the process by which learners come to their understandings. It emphasizes the influence of the pedagogical medium. The excerpt is applied to a localized learning situation drawn from the study, which involved a pair of students investigating the 101 × activity (see Figure 11.1). It demonstrates how their generalizations of the patterns, and their understanding, evolved through interpreting the situation from the perspective of their preconceptions. These perspectives were summoned by personal discourses related to school mathematics, language, the pedagogical medium, and other sociocultural influences. They influenced the manner in which the participants engaged with, and then investigated, the task. However, students' interaction with the task and subsequent reflection shifted their existing viewpoint and repositioned their perspective. The participants then re-engaged with the task from that modified perspective. It was from this cyclical oscillating between the part (the activity) and the whole (their prevailing mathemati-

---

101 Times Table

Investigate the pattern formed by the 101 times table by:
1. Predicting what the answer will be when you multiply numbers by 101.
2. What if you try some 2- and 3-digit numbers? Are you still able to predict?
3. Make some rules that help you predict when you have a 1-, 2-, or 3-digit number. Do they work?
4. What if we used decimals?

Figure 11.1 101 times table task.

cal discourse), with the associated ongoing interpretations, that their understanding emerged.

The students began the task:

> **Clare:** Investigate the pattern formed by the 101 times table. When you multiply numbers by 101.
> **Diane:** Times tables, so we just go like, 2 × that and 3 × that.

Their initial engagement and interpretations were filtered by their preconceptions associated with school mathematics. "Times table" is imbued with connotations for each of them, drawn from their previous experiences. The linking of the term to "multiply numbers" and "2 × that and 3 × that" brings to the fore interpretations of what the task might involve. These position their initial perspectives. Their preconceptions regarding the pedagogical medium were also influential. It was from the viewpoint evoked by these preconceptions that they engaged with the task. They entered the following:

$$101$$
$$202$$

> **Clare:** Yeah, but couldn't we just go times 2 or 101 times?
> **Diane:** Yeah, just do that.
> **Clare:** You go equals, 101 times 2. Then you click in there. Oh, man, we did it. Now what are we going to go up to?

The students' engagement with the task, and the dialogue this evoked, were influenced by their understanding of the situation, the mathematical processes involved (e.g., the patterns), and the pedagogical medium. This interaction had shaped their underlying perspectives in these areas, and they re-engaged with the task from these fresh perspectives.

They re-entered the data, changing the format, to give the following:

A	B	C
101	1	101
101	2	202
101	3	303
101	4	404
101	5	505
...	...	...

Diane: What we did was, we got 101. We went into A1, then we typed in 101. Then we typed in B1, and then we typed in equals A1, then the times sign, then two. Then we put enter and we dragged that little box down the side to the bottom to get all the answers. That gives you the answers when you multiply numbers by 101. We multiplied two by 101. You get 202.

Clare: So, you get the number, zero, then the number again. The next thing is to try other numbers. Like two zero, twenty.

They articulated an informal conjecture for a generalized form of the pattern, based on the visual pattern revealed by the spreadsheet structure, in conjunction with other affordances of the medium (e.g., instant feedback). Once more, their perspective had been modified.

Diane: So, if we do two-digit numbers, can we still predict?

Clare: So, we'll do, like, ten times 101. That's a thousand and ten.

Diane: Shall we try, like, 306?

Clare: No, we'll try thirteen, an unlucky number. That'll be 13, zero, 13.

They enter 13, then drag down:

101	13	1313
101	14	1414
101	15	1515
101	16	1616 etc.

Diane: Wow!

Clare: Cool.

Diane: Now, putting our thinking caps on....

They had anticipated an outcome of 13, zero, 13 (13013) when 13 was entered, consistent with their emerging informal conjecture, yet the output was unexpected (1313). There was a difference between the *expected* and the *actual* output, initiating reflection and a reorientation of their thinking.

Clare: We had the number by itself, then we saw that it was the double. So, with two-digits, you get a double number. What if we had three-digit numbers?

Diane: Let's try 100. That should add two zeros. Yeah, see. OK, now. Now, copy down a bit.

101	100	10100
101	101	10201
101	102	10302
101	103	10403
101	104	10504
101	105	10605
101	106	10706
101	107	10807

Clare: Wow, there's a pattern. You see, you add one to the number, like 102 becomes 103, then you add on the last two numbers [02, which makes the 103, 10302. So 102 was transformed to 10302].

Their engagement with the task, through the pedagogical medium had evoked a shift in their interpretation of the situation. The alternating of their attention from the whole (their underlying perceptions) and the part (the task), as filtered by the pedagogical medium and their interaction, was modifying the viewpoint from which they engaged and the approach with which they engaged the task. It was from their interpretations of this interplay of influences that their understanding was beginning to emerge. This cyclical oscillation from the part to the whole continued, their viewpoint refining with each iteration.

Diane: Yeah, it's like you add one to the hundred and sort of split the number. Try going further.

They dragged the columns down to 119, giving:

101	108	10908
101	109	11009
101	110	11110
101	111	11211
101	112	11312
. . .	. . .	. . .
101	118	11918
101	119	12019

Clare: You see, the pattern carries on. It works.

Diane: Look, there's another pattern as you go down. The second and third digit go 1, 2, 3, up to 18, 19, 20, and the last two go 0, 1, 2, 3, up to 19. It's like you're counting on. Try a few more.

101	120	12120
101	121	12221
101	122	12322
101	123	12423

Clare: Right, our rule is, add one to the number, then add on the last two digits. Like, 123 goes 124, then 23 gets added on the end: 12423. See?

Diane: OK, let's try 200. That should be 20100

They enter 200, getting:

101	200	20200

Diane: Oh . . . it's added on a 2, not a one.

This unexpected outcome evoked a tension with their emerging generalization, instigating reflection and renegotiation of their perspective. The direction of their investigative process shifted slightly. They proposed a new sub-goal or direction to their approach and investigated further.

Clare: Maybe it's doubled it to get 202, then got the two zeros from multiplying by 100. Try another 200 one.

They entered 250, then 251, with the following output:

| 101 | 250 | 25250 |
| 101 | 251 | 25351 |

Diane: No, it is adding two now. See? 250 plus 2 is 252, then the 50 at the end [25250]. Where's that 2 coming from? Is it because it starts with 2 and the others started with 1 [the first digit is a two, as compared to the earlier examples, where the first digit was a one]? See if it adds three when we use 300s.

They entered the following:

| 101 | 300 | 30300 |
| 101 | 350 | 35350 |

Diane: Yes! Now 351 should be 354 and 51, so 35451. Let's see.

The entered 351:

| 101 | 351 | 35451 |

Clare: OK, then will you add 4 for the 400s? Let's see.

They entered some numbers in the four hundreds, producing the following output:

101	400	40400
101	456	46056
101	499	50399

Clare: That last one's a bit weird, going up to a 5.
Diane: It's adding 4, though. See, 499 plus 4 is 503, and then the 99 at the end. Now, how do we put this? It adds the first number to the number, then puts the last two digits at the end. We'll put some more 400s in to see. 490 should be 49490 and 491, 49591. Try.

They entered those two numbers and then dragged down to get the following:

101	490	49490
101	491	49591
101	492	49692
101	493	49793
101	494	49894
101	495	49995
101	496	50096

The participants have negotiated a lingering consensus of the situation—one borne of their evolving interpretations, as they engaged the task from their preconceptions in the associated domains. The ensuing interaction and reflection evoked subsequent shifts in their perspective. They then re-engaged with the task from these modifying perspectives. Each iteration of the hermeneutic circle transformed their interpretation of the situation, with the spreadsheet medium influential to their approach, interpretations, and, inevitably, their consensus of meaning. The mathematical understanding that emerged was inevitably a function of the pedagogical medium employed—in this case, the spreadsheet—and the interplay of their interactions framed their evolving perspectives.

Our perspective influences the sense we make of unfamiliar phenomena. Likewise, the interpretations made by the participants, the researcher, and the readers were influenced by the perspectives they held at that particular juncture, and might have varied in different times. The layering of these local hermeneutic situations informs the macro position, but each retains specificity to its evolution. The data were also indicative of the complexity of influences entailed in a local hermeneutic circle. While the learner, the mathematical task, the pedagogical medium, and the learners' discourses in those and related domains have primacy in the evolution of interpretation and understanding, discourses to do with power, advocacy, and expectation were pervasive. The particular examples employed, the inter-relationships of the particular group, and the manner in which their contributions were fashioned and expressed all influenced the interpretations of and within the process in subtle ways. In the broader picture, even these understated favorings are borne of underlying discourses; that is, everything is brought forth from its interpretative lineage.

Reconciling Technical Aspects and Alternative Forms

The following episode arose from another group's engagement with Rice Mate, a task associated with the doubling of grains of rice for each square of a chessboard. The investigative trajectory was influenced by the pedagogical medium through which the pupils engaged with the mathematical activity. More specifically, the questions evoked, the path they took, and the conjectures they formed and tested were fashioned by visual perturbations—the tension arising in their prevailing discourse by the difference between the expected and actual output. This tension, resulting from working with the spreadsheet environment, promoted the restructuring of their perspective, and they approached the task in a slightly modified manner. The recursion of their attention to the task, and their interpretation through modified perspectives, allowed an understanding of technical and conceptual elements of their activity to evolve. The students began by considering the first square of the chessboard and negotiating a way to double the number of grains of rice in subsequent squares:

> **Tony:** OK, A1*2.

The following output was generated:

A1*2

A1*2

A1*2

A1*2

A1*2

The output was unexpected and related to a technical or formatting aspect. Their mathematical preconceptions possibly enabled them to envisage a sequence of numbers doubling from one in some form, but the unexpected screen output led them to re-evaluate the manner in which they engaged the exploration of the task. Their alternating engagements with the task and subsequent reflection on the output through their mathematical and spreadsheet preconceptions were both facilitating their approach to the task, and the emergence of the technical aspects required to enable that approach.

> **Tony:** In A1, we want 1, and then you go something like, =A1*2, then you go fill down and it times everything by 2. So, 1 by 2, then 2 by 2, then 4 by 2, then 8 by 2, 16 by 2.
>
> **Fran:** To double it? Times 2 more than the one before.
>
> **Tony:** The amount of rice for each year will be in each cell.
>
> **Fran:** What's the first thing we need to start off with? =A1*2?
>
> **Tony:** We have 1 in cell 1 [for one grain of rice], and then we add the formula in cell A2 now.
>
> **Fran:** And then fill down.
>
> **Tony:** Got it. Go right down to find out.

They then entered:

	A
1	1
2	=A1*2

They *Fill Down* from cell A2 to produce the sequence of numbers they anticipated would give them the number of grains of rice for each square of the chessboard. They encountered something unexpected with the following output generated:

	A
1	1
2	2
3	4
4	8
. . .	. . .
. . .	. . .
26	33554432
27	67108864
28	1.34E+08
29	2.68E+08
. . .	. . .
64	9.22E+18

> **Fran:** OK, that isn't supposed to happen.
> **Tony:** 9.22E + 18, that makes a lot of sense [disbelieving].

The output was unforeseen and in a form (scientific form) with which they were not familiar. There was a tension between the expected and actual output, causing them to reflect, adjust their position, and re-interpret. These students initially sought a technical solution to resolve their visual perturbation. They looked for a way to reformat the spreadsheet to alleviate their dubious perceptual position:

> **Fran:** Oh, make bigger cells.
> **Tony:** You can make the cell bigger. Pick it up and move it over.
> **Fran:** That should be enough.
> **Tony:** It still doesn't work.

Still perturbed by what the spreadsheet displayed, they sought intervention, so the notion of scientific form was discussed with them. They indicated that they had a clearer perception of the idea, and proceeded with the task. Tony considered the output 2.25E+15:

> **Tony:** When you get past the 5, you will need a lot of zeros. We'll need thirteen more.

They continued with the task, maintaining the numbers in scientific form as they negotiated a way to sum the column of numerical values. This they managed, drawing on their prior understanding of the technical process required. This generated:

$$1.84467E+19$$

> **Tony:** Yeah! It worked.
> **Fran:** We got it!
> **Tony:** Wow. It's a really, really big number.

Drawing on their freshly modified perspective, they considered how it might appear in decimal notation. However, negotiation was required before they could achieve shared understanding.

> **Tony:** How many zeros?
> **Fran:** 19.
> **Tony:** Did you count these numbers here?
> **Fran:** No.
> **Tony:** You need to count from the decimal point to the end and then add the zeros.

They continued with the task, but carried forward their modified perspective—a perspective moderated through iterations of engagement and interpretation, but initiated by the visual perturbation. Their learning trajectory was shaped, via interpretation and engagement, by the various associated socio-cultural filters, in the case of the spreadsheet environment. Their preconceptions were mediated by the pedagogical medium, and their understanding and explanations had incorporated those modified perceptions.

Particular Ways Actual Learning Trajectories Might Evolve

One of the key aspects of the engagement that was influenced by the spreadsheet as a pedagogical medium was the initial engagement with the tasks. Across a range of activities, the students, sometimes after a brief familiarization of the problem, moved immediately to engage with the spreadsheet environment. Usually, this was to generate tables or columns of data, often through the use of formulas and the *Fill Down* function. This initial engagement allowed them to experiment with the intentions of the tasks and to familiarize themselves with the situation. They more readily moved from initial exploration, through prediction and verification, to the generalization phase. Often, they immediately looked to generalize a formula to model the situation. The visual, tabular structure, coupled with the speed of response facilitated their observation of patterns. Their language reflected this and frequently contained the language of generalization.

The influence of this initial engagement permeated the subsequent ongoing interaction. The distinctive nature of this engagement framed the ongoing interactions, interpretations, and explanations, as the students envisioned their investigation through that particular lens. The actual learning trajectories were shaped by their initial engagement of creating formulas or columns and tables of data to model the mathematical situation.

Digital technologies are generally more conducive to the modeling of mathematical situations than pencil-and-paper media, and the data were illustrative of the ways in which the spreadsheet enhanced this aspect. The capacity to manipulate large amounts of data quickly, coupled with the potential for symbolic, numerical, and visual representations, enabled the students to produce models that could be observed simultaneously. That meant that links and relationships between them could be explored in an interactive manner. As well, when the students were required to relate different representations to each other, they had to engage in activities—such as dialogue, interpretation, and explanation—that enhanced the understanding of both parties.

The pedagogical environment of the spreadsheet was also influential in the generation of sub-goals, as the students' learning trajectories unfolded. As they alternated between attending to the activities from the perspective of their underlying perceptions, and then reflecting on this engagement with consequential modification of their evolving perspectives, they set sub-goals that plotted their ongoing interaction. These were frequently reset in response to the output generated within the spreadsheet environment. Sub-goals were generated, at times, because of opportunities afforded by the particular pedagogical medium. As well as those attributes that facilitated the modeling process, the facility to test immediately and reflect on emerging informal conjectures made it possible for the students' sub-goals to be shaped by the medium. The data demonstrated how the students' interpretations of the situations they encountered were influenced by the visual, tabular structure. The structure allowed more direct comparison of adjacent columns and enabled them more easily to perceive relationships between numerical values on which to base their new sub-goal, often linked to an emerging informal conjecture. It enhanced their ability to perceive relationships and recognize patterns in the data. Seeing the pattern evoked questions. On occasion, the students pondered why the pattern was there, and what was underpinning a particular visual sequence.

While investigating in this environment, the students learned to pose questions and sub-goals. In addition, they were encouraged to create personal explanations—explanations that were often visually referenced, probably due to the pedagogical medium. The environment provided students with opportunities to explore powerful ideas and to explore concepts that they might not otherwise be exposed to. At times, the learning trajectory evolved in unexpected ways (Calder, 2007). When the output varied, sometimes markedly, from what was expected, it created tension that often led to substantial shifts in the way the student interpreted or engaged the situation. This aspect appeared to stimulate discussion. The students wanted to articulate the rapidly generated output and discuss the connections they could see, not least when it was unexpected. This aspect of surprise provoked curiosity and intrigue, allied with the interactive and visual nature of the experience, in the students' general view, made the learning "more fun and interesting." This, in turn, enhanced the motivational aspects of working through the spreadsheet medium—a feature that emerged in the interview, survey, and observational data.

Students' propensity to move beyond known procedures in recognizable situations is indicative of their willingness to try fresh strategies in their approach to investigation and problem solving. By implication, problem solving contains an element of the unknown that requires unraveling and addressing through the application of strategies in new situations or in an unfamiliar manner. This requires a degree of creativity and a willingness to take conceptual or procedural risks of a mathematical nature. It is risk-taking in a positive, creative sense, as compared to risky behavior. The data were indicative of the spreadsheet environment affording learning behaviors and responses that facilitated the learners' willingness to take risks while operating within an investigative cycle. This seemed to allow the students to pose informal conjectures, to explore and then reflect on them, before—perhaps after several investigative iterations—either validating or rejecting them. The posing and investigation of informal conjectures fostered mathematical thinking. These affordances were evident in the spreadsheet environment.

The speed of response to input meant that the spreadsheet was imminently suitable for facilitating mathematical reasoning. When the students observed a pattern or graph rapidly, they developed the freedom to explore variations and, perhaps, with teacher intervention, learned to make conjectures and then pose questions themselves. This facility to immediately test predictions, reflect on outcomes, then make further conjectures not only enhanced the students' ability to solve problems and communicate mathematically, it also developed their logic and reasoning, as they investigated variations or the application of procedures.

The data illustrated ways that the learning experience differed and alternative understandings evolved, through the lens of the digital pedagogical medium. Mathematical thinking was enhanced. This gives credence to teachers making available, and advocating, the use of appropriate digital technology in the classroom and selecting and using tasks appropriate for their use.

THE EVOLUTION OF RESEARCH PERSPECTIVES

The students in this study engaged with the tasks through their preconceptions derived from earlier experiences. Seeing the output of their mathematizing in the visual, tabular form of the spreadsheet modified those preconceptions as they made interpretations of their interaction. In the following brief excerpt, two students were investigating the 101 X activity (see Figure 11.1).

They had produced the following output:

1	101
2	202
3	303
4	404

Tim: So, it's the number, then a zero, and then the number again.
Carl: Yeah, yeah. 5 will be 505. 55 would be 55055. Drag down.

. . .	. . .
13	1313
14	1414
15	1515
16	1616
17	1717
. . .	. . .

Carl: What? It's just repeating.
Tim: Like doubles, so 18 would be eighteen, eighteen and 55 would be fifty-five, fifty-five.

They continued refining their generalization through the modification of their perceptions as they interpreted the outcome of their engagement and adjusted their perspective. Their generalizations were based on the number and positioning of the digits. They used a form of visual reasoning to generalize the pattern (Presmeg, 1986). They then re-engaged with the activity from a fresh perspective, with the interpretation and understanding evolving in this ongoing manner. The broader discourse of mathematics (in this case, visual reasoning) was likewise transformed (albeit slightly) by this engagement. The boundaries of mathematics per se were extended, or existing positions were enriched, by that engagement. Other students commented in the interviews on the way the spreadsheet environment assisted their interpretation. For instance, Chris said, "Columns make it easier— they separated the numbers and stopped you getting muddled. It keeps it in order, helps with ordering and patterns." The cultural formation of mathematics evolved slightly, as the students engaged the tasks through the pedagogical medium of the spreadsheet. Subsequent interpretations influenced the way mathematics was perceived.

The individual engagements of the students also influenced the researchers' perspectives and interpretations of the data, and the research methods that were employed. The analysis of the initial data revealed how the spreadsheet environment might structure the output visually. This analysis, in conjunction with other constitutive influences—for example, the research literature—modified the approach to a more interpretive perspective. Research methods were employed that would give alternative insights into these visual interpretations as the students' attention shifted alternately from preconception to interaction. Viewing the data through this lens gave further insights into the investigation of the research questions—in particular, the ways understanding emerged for the students and the ways the pedagogical medium of the spreadsheet influenced their understanding. The research was modified simultaneously, as research practice drawn from existing prevailing discourses in mathematics education research was engaged for the research process, and as the research process subsequently modified personal perceptions of mathematics education research. Not only was the individual researcher trajectory modified through participation in the research process, it was also transformed as the researcher engaged in collegial dialogue, presented papers at conferences, and published journal articles that focused on the visual tabular structure of the pedagogical medium.

Clearly, the mathematizing at an individual level, the cultural formation of mathematics, the individual research process, and the evolution of mathematics education research are inextricably linked: They are mutually interdependent. They all evolve through cycles of interpretation. Mathematics is an evolving set of perceptions, seeming to become more complex on its peripheries, yet more refined in its core identities, with each iteration of interaction, reflection, and interpretation. The elements of mathematics that are engaged transform the perceptions of the person interacting with the mathematics, but, likewise, those elements, too, are transformed by their engagement with mathematicians, learners, or researchers, even if only by a minuscule amount. The boundaries of mathematics are expanding or becoming more refined through that interaction. Thus, mathematics can be envisaged as a hermeneutic process, one where iterations of engagement, reflection, interpretation, then re-engagement from modified perspectives fashion emerging ideas.

Further Pedagogical Implications

The spreadsheet environment reshaped the students' approaches and the manner in which they traversed their actual learning trajectories, by the particular nature of their experiences while working within that environment. It allowed them to engage in alternative processes and to envisage their interpretations and explanations from fresh perspectives. The mathematizing facilitated by the medium was transformed by the visual, interactive nature of the investigative process. Students used visual elements in their reasoning, while their explanations were punctuated with visual referents, such as the position and visual pattern of the digits. As such, the generalizations that emerged were couched in visual terms. They interpreted and explained their reasoning in alternative ways. Importantly, the visual tabular structure enhanced the possibility of seeing relationships in ways that might otherwise have been unattainable or inaccessible. Coupled with other affordances, such as the increased speed of the feedback, this visual dimension expanded the boundaries of what constituted mathematical knowledge, and gave students access to ideas earlier than teachers' usual expectation. It allowed a shift in focus from calculation techniques to a focus on mathematical thinking and understanding. Armed with the capacity to think mathematically and to generalize, enhanced by the simultaneous viewing and translation between the alternative forms, the learners' modeling of the situations with various representations fostered the reorganization of their thinking. This phenomenon extended the scope of mathematical tasks with which students engaged, because they linked visual, numerical, and symbolic representations of the data. In statistics, for example, this permits a more exploratory approach to data analysis, while the interactive and rapid nature of organizing data promotes the posing of statistical questions and an immediate exploration of those questions.

There are other pedagogical implications associated with the alternative understandings that emerge through the gaze of digital pedagogical media. The range of investigative tasks can expand to include those where the speed and precision of multiple computations would otherwise be restrictive. This opens up the teacher's repertoire to include more real-life situations that otherwise produce vast ranges of "untidy" numbers, and are difficult to manage in a pencil and paper environment. While dynamic visual imaging opens opportunities for learning, there are also pedagogical considerations associated with the nature and number of visual images produced. Some researchers warn that the "noise" created by greater exposure to visual images will constrain students' capacity to mentally image (Mason, 2005). However, the greater propensity to investigate demonstrated by the students in this study fosters the possibility of a more child-centered approach. Students might more easily negotiate, research, and explore their own questions, and pursue more individual, self-monitored learning trajectories. This raises other questions related to pedagogy. How might the teacher's role become more of a facilitator and monitor of students' learning experiences? How might teachers maintain participation across a range of abilities and learning contexts?

Perhaps more integrated curriculum with student-negotiated investigation of meaningful problem situations will be possible, given that digital media permits more fine-grained recording and monitoring of mathematical elements. The potential for students to research information and processes on the internet enhances their potential to pursue individual investigation, while also permitting the formation of informal investigative networks beyond the school's boundaries. How the classroom, the movement within it, and the social structures might evolve in more informal working situations involving small group and individual investigative study is a question that needs to be examined. Teachers, schools, and communities would need to be comfortable with this alternative learning model, and would require different expectations about how the development of mathematical thinking might manifest in a classroom.

On a broader level, the impact of digital technologies on society and investigative processes, in general, offers scope for changing the nature of some elements of mathematics and mathematical thinking. While there is recognition in some quarters within the mathematics community that some evolution has already occurred (for instance, the emergence of visual reasoning as a "legitimate" form of mathematizing), there is certainly no consensus within that community regarding this aspect, nor orchestrated intention to explore the boundaries of such possibilities. In the domain of mathematics education, digital technologies are given greater privilege, although their potential use in the classroom is still only partially realized. Modeling is one aspect of mathematics education that might be given greater primacy in pedagogical practice. The nature and immediacy of feedback, which was featured in the analysis, enables the successive refinement of informal conjectures and solutions. It would seem that pedagogical approaches in mathematics education will need to evolve to allow greater access to digital technologies, and to enhance the opportunities for students to engage with mathematical investigation and thinking those investigations through.

REFERENCES

Arzarello, F., Paola, D., & Robutti, O. (2006). Curricula innovation: An example of a learning environment integrated with technology. In C. Hoyles, J-B Lagrange, L.H. Son, & N. Sinclair (Eds.), *Proceedings of 17th ICMI Study conference, Technology Revisited.* Hanoi: Hanoi University of Technology.

Brown, T. (2001). *Mathematics education and language: Interpreting hermeneutics and post-structuralism*. Dordrecht: Kluwer Academic Publishers.

Brown, T. (2008a). Lacan, subjectivity and the task of mathematics education research. *Educational Studies in Mathematics, 68*, 227–245.

Brown, T. (2008b). Signifying "learner," "teacher," and "mathematics": A response to a special issue. *Educational Studies in Mathematics, 69*, 249–263.

Calder, N. S. (2007). *Visual perturbances* in pedagogical media. In J. Watson & K. Beswick (Eds.), *Mathematics: Essential tools, essential practice*. Proceedings of the 30th annual conference of the Mathematics Education Research Group of Australasia, Hobart (Vol. 1, pp. 172–181). Hobart: MERGA.

Calder, N. S., Brown, T., Hanley, U., & Darby, S. (2006). Forming conjectures within a spreadsheet environment. *Mathematics Education Research Journal, 18*(3), 100–116.

Confrey, J., & Kazak, S. (2006). A 30-year reflection on constructivism. In A. Gutierrez & P. Boero (Eds.), *Handbook of research on the psychology of mathematics education past, present and future*. Rotterdam: Sense Publishers.

Gadamer, H. G. (1989). *Truth and method* (Trans: J. Weinsheimer & D. G. Marshall). NY: Crossroad Press.

Lerman, S. (2006). Socio-cultural research in PME. In A. Gutierrez & P. Boero (Eds.), *Handbook of research on the psychology of mathematics education: Past, present and future* (pp. 347–366). Rotterdam: Sense Publishers.

Marriotti, M. A. (2006). New artifacts and the mediation of mathematical meanings. In C. Hoyles, J-B Lagrange, L. H. Son, & N. Sinclair (Eds.), *Proceedings of 17th ICMI Study conference, Technology Revisited*. Hanoi: Hanoi University of Technology.

Mason, J. (2002). *Researching your own practice: The discipline of noticing*. London: Routledge Falmer.

Mason, J. (2005). Mediating mathematical thinking with e-screens. In S. Johnston-Wilder & D. Pimm (Eds.), *Teaching Secondary Mathematics with ICT* (pp. 81–100). Berkshire, UK: Open University Press.

Presmeg, N. (1986). Visualization in mathematics giftedness. *Educational Studies in Mathematics, 17*(3), 297–311.

Radford, L., Bardini, C., & Sabena, C. (2007). Perceiving the general: The multisemiotic dimension of students' algebraic activity. *Journal for Research in Mathematics Education, 38*(5), 507–530.

Ricoeur, P. (1981). *Hermeneutics and the human sciences*. Cambridge: Cambridge University Press.

Walshaw, M. (2001). Engaging postpositivist perspectives in mathematics education. *SAMEpapers* 2001, 19–33.

FURTHER READING

Johnston-Wilder, S., & Pimm, D. (Eds.). (2005). *Teaching secondary mathematics with ICT*. Berkshire, UK: Open University Press.

Hoyles, C., & Lagrange, J-B. (Eds.). (2009). *Mathematics education and technology–Rethinking the terrain. The 17th ICMI study*. New York: Springer.

CHAPTER 12

THE PARADOX AND POLITICS OF DISADVANTAGE

Narrativizing Critical Moments of Discourse and Pedagogy

Dalene Swanson

ABSTRACT

Amartya Sen (1999), winner of the Nobel Prize in Economics, wrote in *Development as Freedom* that "focusing on human freedoms contrasts with narrower views of development, such as identifying development with the growth of gross national product, or with the rise of personal incomes, or with industrialization, or with technological advance, or with social modernization" (p. 3). In attempting to refocus the intent and purpose of social development away from how it has been characterized in contemporary times, Sen holds us to account, ethically, for our participation in a globalizing project in which we are all implicated, and yet whose effects we do not fully appreciate. Such "effects" are ideological, in that they operate at the level of structures and interstices, and are encoded within complicities of normative systems of human relations. They are, consequently, inherently moral. What are the implications for education, in general, and mathematics education, in particular, when industrialization and economic growth are the foremost policy objectives of a nation state? I will explore that

Unpacking Pedagogy: New Perspectives for Mathematics Classrooms, pages 245–263
Copyright © 2010 by Information Age Publishing
245

issue in this chapter from a philosophical standpoint and apply the issue to the development of mathematics curriculum in a setting within South Africa.

For Sen, human rights and freedoms of all kinds should lead technological, scientific, and economic "progress," rather than become pleasant side effects of these "development instruments." In his view, human rights and social and political freedoms (and I would include ecological rights) are the necessary constituent components of development. As Sen would maintain, capitalism's rapid spread has been at the expense of much, as well as many who are often left marginalized and voiceless in the process. Capitalism has failed to provide the alluring "rewards" for millions of people living in abject poverty who have little agency in relation to the hierarchy of access it has produced and which it serves to reproduce.

The assumption that scientific and economic advancement will necessarily *achieve* the objectives of liberty and justice "freely" or "naturally" is necessarily "dangerous." However, our most widely shared understandings about what constitutes the "common good" for all humanity are constructed on the basis of this assumption. Importantly, the political complexity of the association between modern scientific industrialization and that of human and environmental wellbeing is often lost. Troubling the ideological underpinnings of the large-scale development model leads us to question what this "advancement" is towards, how it is "better" (and for whom), what is lost, what is gained, at what price, and what the ends are. From a pedagogical perspective, some important questions follow: What implications does the global model of development have for the design of curricula, for the creation of educational structures and environments, for ways of choosing to engage with and construct participants within learning communities, and to what ends? Is justice a nicety to or necessity of education? Does it have an intrinsic or extrinsic relationship to it?

Economic development has become the ongoing hegemonic mechanism that sustains the neo-colonial project on a global(izing) scale. Ideological assumptions underwriting the global project of development operate to monitor, inform, and normalize accepted progressive discourses on education (Skovsmose & Valero, 2001). They measure and control meanings on how success and failure is constituted, on what is valued as knowledge, on who has access, on who "they" are, on what constitutes good educational practice, on what the aims are for educational practice locally and globally. I argue that the way in which "success" and "disadvantage" in educational contexts is constituted is critically related to the social identities through which participants are differentially positioned (Swanson, 1998, 2001, 2004, 2005, 2006). This critical relationship is most dominantly informed by specific assumptions about what is "good" for "a learner." This "learner," within the educational stream of our progressive modernization project, is someone who most likely has been demographied and psychologized, but also, ironically, has been dehistoricized and decontextualized from community and environment.

Conceptualizations of "the learner" are most often driven by dominant Western educational discourses that normalize competition and draw on individualistic ideological investments globally. These prevailing discourses enable life opportunities for individuals within certain valued groups, while delimiting opportunities for others. In so doing, they reify dominant cultural formations over localized ones, and these dominant discourses become the master print for entry or denial of access. Life opportunities are, however, beyond a question, of mere "access." Normalized assumptions inhabit questions of what is valued, what is conserved, and what is foreclosed, in terms of being and imagining within other frames of reference. The ways in which these ideological assumptions impact on the recognition and validation of indigenous, generational, or localized ways of knowing and being, and how they permit or enclose imaginative possibilities for communities to be otherwise, are all interconnected and relate directly to the (false) promise of the ends of freedom and (misconception of) wellbeing through the instrumental and material means of techno-scientific and economic "progress."

The promise of greater "wealth" and national strength through participation in global competition is thwarted by endemic power differentials between nations and populations, situated within hierarchical "social divisions of labor." Thus, capitalism could be seen to reveal a "paradoxical and schizophrenic" nature (see Deleuze and Guattari, 2005) in its reliance on the "winner and loser" game-playing that is recontextualized (Bernstein, 2000; Dowling, 1998) at the level of community. It is in that sense that capitalism is said to perform a "symbolic violence" (Bourdieu & Wacquant, 1992) on racialized, gendered and socio-psychologically pathologized bodies (such as socially constructed learners) within various educational contexts (such as classrooms), discourses, and practices (of, for example, school mathematics) in diverse locations across the world. Increasing neo-liberalization of institutions and the global modernization agenda has set the terms of global economic and social participation by increasing the monitoring and regulation of individuals, groups, and targeted communities. Such measures serve to perpetuate the global neo-colonial project. The current conception of development, framed, as it is, as "economic progress" within the neo-colonial project, has become a "Truth" that tolerates little resistance, that excludes a range of other possible meanings and ways of engagement, and that attempts to silence alternative voices. The more discourses on development become increasingly foreclosed, in these terms, the greater freedom and the possibilities of freedom become enclosed.

INTRODUCING IDEAS

Mathematics as Implicated in the Global Project of Development

Mathematics is deeply implicated in the global project of development. Such a realization draws our attention to the strong techno-scientific utilitarian pull that the mathematical sciences afford and to their preeminence in the "division of labor" (Bernstein, 2000) in the social domain. The power of the voice of the mathematical sciences is such that it casts a "mythologizing gaze" (Dowling, 1998; 2001) that "recontextualizes" (Bernstein; Dowling) the social, cultural, historical, and political contingencies and complexities of localized settings of mathematics education practices to the "regulating principles" (Bernstein, 2000) of Western mathematical discourses. This is achieved via its mythologizing reference to the redemptive powers of techno-scientific "progress" as the "savior" of humanity and of the environment. Rationalist Enlightenment, instrumental in technological and scientific achievements, sustains an imagination of the West as the "rightful" supremacy. In doing so, it makes possible a discourse of "rescue and redemption" in relation to marginalized groups and individuals. This is very often the case in "developing" contexts of the world. These "rescue and redemptive" discourses are, for example, very much at play in many African contexts. They construct African people as "vulnerable others" that need to be "uplifted" and "saved."

The mythologizing and recontextualizing gaze (Dowling, 1998; 2001) of the mathematical sciences is implicated both in the development of constructions of pedagogy and in the development of mathematical identities. Mathematical literacy goes to the core of one's citizenship in the individualistic neo-liberal parlance of Western mathematics curricular documents (see Swanson, 2008). Within the constructions of citizen made possible by the nation state, economic capacity is closely tied to "levels of literacy," and, in particular, to levels of scientific literacy. Mathematical "failure" is thus framed within discourses of need for the nation state. In this thinking, mathematics education is in dissonance with democracy (Skovsmose & Valero, 2001). Thus, the constructed "failure" in mathematics is inextricably informed by citizenship; intersubjective social difference discourses on race, gender, socio-economic class, ethnicity, ability, and culture; and other geopolitical differences. Zevenbergen (2003), Lerman and Tsatsaroni (1998), and Dowling (1998) have provided useful discussions of how "failure" is constituted within mathematical discourses, performing "violence" on bodies, in the context of its production and in the context of the identities it constitutes. My own work has also elaborated on this issue (Swanson, 1998, 2000, 2002, 2004, 2005, 2006). For example, in the following excerpt, in a response to Eric Gutstein (2008), I address the question of mathematics and citizenship, and the oppressions this relationship generates:

> Common progressivist and utilitarian rhetoric on the "importance" of mathematics learning in schools often make claims to "good citizenship" and vocational advancement. A "successful citizen," according to this tenet, is one that has access to the power of mathematics to "know the world." ...Yet, the politics of such "coming to know" is most commonly denied, so that mathematics' ability to enable its knowing subjects to "describe our world" is purportedly di-

vorced from subjective influence and human interference: Mathematics has great utilitarian worth here, but is untainted by the messiness of politics and human vulnerability. "Failure," in these terms, is therefore constructed, ironically, as a condition of being an unknowing mathematical subject....

... Standardized testing, streaming/tracking systems in schools for mathematics and pronounced differentiated teaching practices in this subject, as well as other gate-keeping controls, ensure that a differentiated hierarchy of access is produced that emulates, assists, (re)produces, and is (re)produced by the hierarchy within capitalist relations of production. Mathematics' high status in the "social division of labour of discourses" (Bernstein, 2000) within schools and society, makes it a high stakes game to play, and its "strong grammar" (Bernstein, 2000) provides it with significant cultural caché for those with the luck and privilege to have access to it as knowing subjects and citizens. (Swanson, 2008, pp. 213–214)

The often suppressed ideological investments of mathematical discourse in various contexts align strongly with questions of ethical engagement. Questions concerning why we "do" or "don't do" mathematics, who has access, and what kinds of mathematics are taught or not taught to whom, and why, are caught up in these considerations. The current paternalistic, savior-of-Africa paradigm, supported by a "cultural politics of benevolence" (Jefferess, 2008), and framed within a neoliberal and neocolonial economic development model, aligns, similarly, in terms of investments of power and interest, with the way in which mathematics curricula are constituted within citizenship discourses that feed off nationalisms and competitive globalization agendas. In this way, such curricula perpetuate new techno-scientific hegemonies and industrialization that advance neocolonialism.

It is my contention that the power that mathematics asserts within the social domain is divisive and extensive, and serves to rationalize and normalize cultural, socio-historical, and geopolitical differences and inequalities, rather than complicate or disrupt them. A critical awareness of the operationalization of such networks of influence, their critical interconnections with discourses of power, the deployment of ideologies, and the way in which social hegemonies are produced on a global structural level, are aspects of the political economy of mathematics education that are, for the most part, ignored in academic discussions. Unless there is a fuller appreciation of these complex issues, there is, I believe, little chance of cultivating a deeper understanding of how one might approach mathematics education as a moral and ethical practice. At the core of any pedagogical practice should lie the moral questions of "why" and "for what" and "for whom" and "to whom" and "with whom" are we teaching this and, if so, "Why are we teaching this way?" It should also be asked: "With what and whose wisdom? What investments of power and ideology lie hidden in any single judgment to act mathematically and/or pedagogically in a chosen way? What assumptions about what is good for those engaged in mathematics education precede us without our awareness to question or to choose to act otherwise?"

These are complex issues, and they deserve due diligence, dialogue, and contemplation. I propose, somewhat tentatively, a few ideas about how we might approach the oppressions and hegemonies noted earlier. It is dangerous and unwise to focus at the micro level only. Any such focus needs to be framed within the context of macro discourses at the structural level, and possibly on a global scale. Such a broad focus is crucial if we are to maintain an awareness of how dominant discourses in the social domain circulate to reproduce and sustain or change discourses and limit or enable possibilities for individuals or groups. In a paradoxical Foucauldian sense (1981), discourse permits as it dis(en)ables; it produces subjectivity as much as it allows us to thwart the oppressions it reproduces. Hegemonies of mathematics education work alongside discourses of economic development, neoliberalism, neocolonialism, nationalisms, and social divisions, such as race, ethnicity, gender, geography, ability, language, socio-economic status, sexuality, religion, cult, and otherwise. Isolating any one of these issues would force us to lose a sense of their interrelatedness, complexity and the intricacies of power and influence. Keeping this in mind, I nevertheless make a case for a critical, "culturally-conscious" theory of curriculum as one beginning pathway into possibilities for rethinking egalitarianism, democracy, and justice, in respect to a critical mathematics education. In a sense, I have already engaged, to some degree, with a critical, culturally-conscious approach through my earlier critiques by viewing mathematics education as a political (and moral) text. Such an approach can be viewed as a curriculum approach in acknowledging the work within the reconceptualization movement (see Pinar, Reynolds, Slattery, & Taubman, 1996). In advancing a critical, culturally-conscious theory of curriculum, it is necessary to understand the work of theories of curriculum, and how they might help us in our approach toward a more purposeful and ethical engagement with mathematics education. Henderson and Kesson (2004) draw on Walker (2003) to argue:

> Curriculum theories...are about ideals, values, and priorities. They employ reason and evidence, but in the service of passion. Curriculum theories can be analytical as well as partisan, but unlike scientific theories, they are not curriculum theories unless they are about ideals. Curriculum theories make ideals explicit, clarify them, work out their consequences for curriculum practice, compare them to other ideals, and justify or criticize them. (p. 60, p. xiv)

To open the debate, what curriculum, then, should be undertaken in the mathematics education context to support *an ideal*, and *which* ideal? If *the ideal* centers around a purpose of/for education, as Gert Biesta (2009) reminds us, then it might well be advisable to look at issues of democracy, justice, and egalitarianism as offering worthy purpose, even a "critical responsibility," in the Freirian (1999) sense, through conscientization, as long as this is not through a singularly anthropomorphic lens of social justice that is carried out narrowly and unholistically at the expense of urgently important ecological justice considerations (Bowers, 2001; Bowers and Apffel-Marglin, 2005). Bishop (1990), amongst others, has written of the hegemony of mathematics education as a form of Western imperialism. Powell and Frankenstein (1997), and others, have addressed the issues of Eurocentricism in dominant mathematics education practices throughout the world. From a strongly sociological position, different, in many ways, from the perspectives of Bishop and that of Powell and Frankenstein, Dowling (1998) has also addressed the power of mathematics and its divisiveness through its various discursive elaborations in contexts. Skovsmose and Valero (2001) address mathematics education in terms of its critical relationship with democracy and the need for a transformative pedagogy. "Mathematics for Social Justice" has also gained momentum (see Gutstein, 2008), while D'Ambrosio (1997, 2001, 2006) has inspired the Ethnomathematics movement as an approach that might be used for resisting the hegemony of Western mathematics education. Specifically, ethnomathematics is said to lie "on the borderline between the history of mathematics and cultural anthropology" (D'Ambrosio, 1997, p. 13). Like ethnomathematics, "Everyday Mathematics" is said to exist in the social habits of people throughout the world. While they are not without ambiguities and problems, these approaches offer valuable insights and practical approaches that highlight the cultural investment that other approaches often oppressively endorse as "culturally neutral" and objective. Such "culturally conscious" approaches, even as they are fraught with contradiction and paradox (see Dowling, 1998, for a critique of Ethnomathematics), nevertheless, draw attention to the implicatedness of Western mathematics in divisive social structures. In these structures, economic and ideological models become hidden curricula.

Another potentially promising approach that reflects a cultural consciousness is a "culturally-responsive" mathematics education. Drawing on Ethnomathematical thinking and aspiration, it bears witness to mathematical practices as value-laden (Bishop, 2001) and to relationships within classrooms as culturally-informed. There is a sense of reciprocity—a sense of cultural relationality, in terms of what is learned and how it is constituted. It is within this line of thinking that Barta and Bremner (2009) recognize that "mathematics in practice becomes an issue of identity as well as cognitive process" (p. 91). They note that "[s]ome aspects of culturally embedded mathematical knowledge are amenable to formal description, but do not fully capture the mathematical competencies of the people who use that knowledge" (p. 91). These competencies are drawn from the social, physical, historical, as well as political contexts in which various groups live. The paradox of culturally responsive pedagogy and its double-bind, however, lies in the very hegemony that it wishes to address. Children of the First Nations of Canada's indigenous groups, and other marginalized indigenous groups throughout the world, need to be constituted as "other," as different, to support an approach whose ultimate objective is for emancipation, democracy, and egalitarianism. Donald (2009) speaks of these "civilizational frontiers" as "unquestioned" and "a naturalized idiosyncrasy of Canadian society" (p. 23). Such an idiosyncrasy maintains and reproduces the "socio-spatial separation of Canadian (insiders) and Aboriginal (outsiders)" (p. 23) that need to be "included" or "responded to," or that the insiders (as the controllers of educational discourse, curriculum and policy) need "to be responsive to." As Donald notes, "Aboriginal peoples and their ways have been reduced to an existence *outside* of Euro-Western civilization" (ibid.). The difficulty is that, for any group of marginalized peoples, "cultural difference" is inevitably a socio-historical construction. What becomes "relevant" for these students to learn based on conceptions of their culture is also somewhat problematic. Who decides what is relevant to learn for whom, and how do these decisions position them as learners or knowers in the social domain within prevailing power relations and discourses of difference?

Each curriculum approach, including culturally-conscious ones, to mathematics education is fraught with contradiction, ambiguity, and paradox. Without a critical consciousness and language of resistance, we may be forced to unwittingly rely on uncontestable assumptions as truth that we are disempowered to contest. Because of the significant power of influence that mathematics commands in the social domain, it is even more important to develop a language of empowerment to critique, question, and contest such influence where it does not serve the ends of worthy human ideals, ones that encompass social justice and democracy. Noting and acknowledging the way in which mathematics education is implicated in socio-historical and contemporary investments of power, geopolitically, and recognizing the ideologies it supports and operationalizes, and the oppressions it sustains, is crucial to the development of a disposition that seeks more democratic and more ethically just curriculum approaches to mathematics education.

In the following section, I provide situated examples of lived experience of the potentially oppressive nature of mathematical discourse in a political context. The examples offer a narrative rendering of the issues of ideology and ethical obligation. In the first excerpt, the context is a historical missionary school in a township or informal settlement in the Cape Province of South Africa, a community in which I engaged in my inquiry. The excerpt highlights ethical dilemmas and exemplifies ideologically-informed assumptions about what is "good" for an impoverished community, both mathematically and politically speaking. The excerpt is drawn from a narrative piece, *Fishes and Loaves: A Parable of "Failure."* The use of biblical language, therefore, is a metaphorical elaboration of the parable, and is commensurate with the Christian context of the local schooling community.

This excerpt is drawn from my doctoral research (Swanson, 2004) in a postapartheid context. Another version of parts of this excerpt appear in Swanson (2007a). Similar in much of its purposes, sentiments, and effect to *narrative inquiry* (see Connelly & Clandinin, 1990, 1991), I developed a methodological approach to my research, which I most often refer to as "critical narrative engagement" or "reflexive rhizomatic narrative." Through "moments of articulation" (Swanson, 2004), greater critical rhizomatic depth becomes possible through such a methodological approach that exceeds the boundaries of a more canalized, formally structured research project. It is flagrantly subjective, in recognizing the impossibility of "objectivity." It aims for authenticity, while recognizing the socially-constructed nature of "truth." Unlike more positivist qualitative research approaches, critical narrative methodologies are more able to capture nuance, ambiguity and contradiction, and address dilemma, the unexpected, the ethically fraught, while captivating audience and offering ethically-charged insights that are of a deeply human, heartfelt, and soulful nature. Rather than seeking packaged "solutions," as in many other scientistic qualitative research approaches that often further neocolonial methodological approaches, narrative methodologies emphasize reflection and contemplativeness. They aesthetically and insightfully seek to bring the personal into the political, connecting the momentary and localized with the broader social domain, the universal, and the global. Deep levels of reflexive engagement become possible with critical narrative rendering as a research objective.

APPLYING IDEAS

A Narrative Unfolding: Implications and Implicatedness of Practice

And so it came to be that I found myself in the midst of "the multitudes," a class of fifty grade 7 children from this informal settlement school. Their teacher had *already* abandoned them for more than three weeks, but they came to school, nonetheless. And I can only assert that they were compelled to come, NOT by the promise of pedagogic empowerment, because the paucity (or non-existence) of subject-based knowledge mitigated against this, BUT by a sense of commonality, of community, and the knowledge of a "place of belonging."

Behind the skull of Apartheid lurk vestiges of the old, so-called Bantu Education, an "impoverished" form of the *already* limited Christian National Education with which we were all indoctrinated, as children of the Apartheid state. Bantu education was imposed on black African and so-called "colored" children . . . the future "hewers of wood and drawers of water," as the Nationalist government liked to refer to Black labor as in those days. This biblical reference of woodcutters and water carriers was, at that time, a hallmark of an ideology, which viewed black African people as inferior and only capable of menial labor . . . and *the legacy of the system remains.*

"Would you like me to teach you some mathematics?" I offered. "Ja, asseblief, mevrou! Ons sal baie daarvan hou! Ja, asseblief, mevrou!" ("Yes, please, ma'am! We would like that very much!"). They began to dance in their desks with excitement at the prospect of learning something . . . something *new* perhaps?, perhaps learning something *differently*?, learning something from *me*?, or perhaps just learning *something . . . anything.* . . . I was moved and heartened, and I began to bless and break the bread of my mathematical knowledge, my own empowerment, and divide it with affection and compassion, and I broke of this body to give of the light and joy of this subject I loved so much, offering it in tasty morsels. This was *surely more* than mere fishes and loaves!!

And I saw those glimmers of light, the kindled glow turn from inward to outward, and flickers of understanding pass across the intent faces of these psychologically-abandoned, pedagogically-abandoned children. And after a while, the children began to answer my questions and even *to ask* questions and participate in the discussion, giving meaning through their bodies, giving back unsparingly of their enthusiasm. I was greatly heartened, as I saw this as tremendous "advancement" in such a short time. For children that I had witnessed as having been exposed to nothing but transmission, rote-learning, and proceduralism (on the occasions when they were exposed to subject learning in the classroom at all), this was an "opening of minds," an "awakening of spirits," a "pedagogic achievement," a "progressivist success."

I was elated, ecstatic! I was performing a miracle. I was proving that the miracle was possible, that my miracle could set a spark in the dry veldt of despair and disillusionment and would Light the Dark and heal my whole country with a Sanctifying Fire. And just when we were about to consecrate the communion of Mathematical Thought, there was a Divine Visitation. The door swung open and a child entered. He handed me a crumpled white bag, and was gone as suddenly as he had come.

A cloud passed over the sun and, through the broken panes of the classroom window, the streaks of golden sunlight dulled and disappeared. The atmosphere cooled; the mood of the children changed. . . . And then there seemed to be a movement, indiscernible at first, and then ever increasing, a spiraling force drawing the atmosphere inwards, like a vortex, deep, downwards into what I was holding, *a crumpled plastic bag*! The children began to move around in their desks in agitation. They were no longer focusing on the mathematics we had been doing—just the bag in my hands. The moment of Mathematical Mastery, of Conceptual Glory, was shattered!

At that moment, I did not know what was happening around me. *I* was now the one without immediate understanding, although, on a deeper level, having grown up in Apartheid South Africa, I recognized this as a "possibility of context" all too well! Nevertheless, I must have shown shock and confusion in my face. "Dis die Appeal, mevrou! Dis die Skool Appeal, dis ons kos van die Appeal af!" ("It is the Appeal, ma'am! The School Appeal. It is our food from the Appeal!") they let me know, moving from their desks in an agitated dance toward me, toward the bag. "Watter 'Appeal' is hierdie?" ("What 'Appeal' is this?") I asked in confusion. And they told me in Afrikaans: "It is the white people, ma'am, that give us our school lunch. It comes from the children in the privileged schools, ma'am. It is for us, ma'am." Their hands began to touch on the sides of the open bag, to touch my hands, to look inside the bag. Was there enough food today, perhaps? No, never enough! I looked into the bag and saw a few sandwiches and fruit...white children's discarded lunches that had been collected for the day and brought to the school under the guise of "assistance" from the surrounding community. "Asseblief, mevrou, gee vir my. Ek is so baie honger vandag!" ("Please, ma'am, give it to me. I am so hungry today!"), they told me, competing with each other to gain my attention or to catch my eye, so that I might take pity on them over the next. I realized that I had the impossible task of having to decide who would eat and who go hungry that day. Everything had seemed to change...or had it? I had offered to teach these children mathematics; now I was expected to preside as judge and jury over their bodies. I was no "liberator" or Great Redeemer, but an *accomplice* coerced into the discourse and practice of Oppression!

The rules of the discourse of mathematics had shifted to a new discourse, whose dominant and uncontested rules won the day. Instead of providing these children with empowerment through access to the "regulating principles" of school mathematics, I was trying to bricolage some moments of pedagogic meaning, draw some understanding from the context to enable a way forward; impossible! I realized with an Illuminating Light, that I was *no* Messiah. I could not provide the miracle of "fishes and loaves." Just as I was not able to perform it pedagogically, so I could not break up the sandwiches and divide the fruit equitably among 50 children so that they all could be satiated. Some would have to starve, and who would those be?

For a fleeting moment, I heard the voice of the progressive mathematics educator: "Draw on the life experiences of the children to help them concretize their mathematical thinking and see relationships between mathematics and real life, to see the relevance of mathematics to the real world." In this context, under these circumstances, what utter useless rhetorical nonsense! The children already *knew* that the principles of divisibility would not work here, just as I knew my inadequacy in providing the Miracle of Divine Multiplication.

White chalk dust from my fingers billowed in a fine mist as the movement of small black hands over mine disturbed it. For a moment, it clouded the view of the contents of the bag, and I thought I saw through the mist the *skull of my country* looking back at me, and in it was *my own skull.* I had tried to provide a skin over that skull, to give it substance and embodiment through my own mathematical empowerment in a context where pedagogic possibilities are reduced to the rules of "poverty." What did I think I was going to do? What Messiah did I think I was? Was I going to "uplift" this community, provide their children with the pedagogic promise of something "better" than fishes and loaves? What "good" did the patronizing offer of food for "disadvantaged learners" do for this community's educational, political, and socio-economic empowerment? In what way did my actions or those of the other do-gooders address the structural and material conditions of the lives of the children and people of this settlement community?

I began to divide out the fare in the classroom, desperately trying to find some rule of fairness to apply to an unjust task, ever aware that the broader injustice lay outside of the classroom, impinging on it. . . . The school classroom, intentioned as a place of *pedagogic empowerment,* became a place of *pedagogic impoverishment,* and one where the throttling rules of poverty reproduced themselves and were well learned and established!

At the same time, in another, very different community school a few kilometers away, children were learning mathematics with a breathless urgency! "Die kleinjies moet eers kry," I said ("The little ones must be offered first."). It was all I could think of. Was I trying to salve my own conscience because I could not find a fairer rule? Those respondents to the School Appeal who had donated the lunches, did they salve their consciences for the day? Could they see inside the classroom and view how their neo-liberal actions had played out? How teaching had been interrupted to satisfy more immediate needs in ways that reinforced dependency and held these people to their poverty? Had this helped to *uplift* a "disadvantaged community," or *establish* it? Was it facilitating Africanisation and empowerment? Or was it merely "fishes and loaves," a parable of "failure"?

Roots/Routes

In the second reflexive narrative, *Roots/Routes,* I attend to the "construction of disadvantage" and the perpetuation of deficit discourses in contexts of constructed and lived poverty. This excerpt is drawn from my doctoral dissertation, and a version of it appears in Swanson (2007b), and another in Swanson (2009). In this excerpt, through a critical moment of research engagement, I address the "blame" paradigm and its unethical investment in oppressive pedagogies. Caught in the dictates of the progress model, identities are framed in situated contexts by the dominant discourses from the social domain that perform symbolic violence, delimiting opportunities for transcendence. I attend to my own complicity in the ideological assumptions of this project and my perpetuation of the oppressive discourses it sustains.

In the same informal settlement school, I am speaking to the principal in his office. In the words I use in the narrative, I want to ask him why it is that, from my perspective, I have not *really* evidenced any real attempt to engage in any progressive education practices within these classrooms; why I have seen *so much* rote learning, when any pedagogic learning took place at all; or why I have seen, from my perspective, so much apparent indifference; why it is that corporal punishment is still used here, when it has been made *illegal* to engage in physically punitive practices in South African schools; why so many of the teachers are so seldom *in* the classroom, when the then National Minister of Education has made urgent and repeated appeals to teachers across the country to take their jobs seriously, for the country's sake, for the sake of our youth and the future generation of South Africa now in creation? Where does the proverbial "buck stop," who is responsible, who cares, why not, and how can we make a difference? I want to ask him why he closes the school early so frequently, causing very small children to have to walk home alone, often unescorted, back to their homes in the informal settlement where they are not attended to or protected because their parents or caretakers are at work? Where does his responsibility to the community end or where does it start? Why does he use class time to have meetings with his staff, and why so frequently is learning interrupted for, apparently, *from my perspective*, inconsequential issues? Why does he legitimize teachers' missing classes by engaging in these practices himself? Why can't meetings take place after school?

As I ask these questions in my mind, I am angry, and my mental discourse is one of judgment and blame. At that moment, a bullying incident occurs in the playground outside the window of the principal's office that helps me attend to my own perspective and reveal the source of the "blame paradigm." I continue:

At that moment . . . and it was not an epiphany, but a slow, blurred form taking root . . . re-rooting in my mind. It was a slow re-realization of what I had done by wanting to "speak out" and to tell this principal that I thought it was "just not good enough." . . . It was a re-cognition of my *own* voice of violence of what brutality I had done in feeding into the discourse on "disadvantage." I re-realized that my thoughts, framed within the discursive roots of my socialization, my education and knowledge, my own perceived empowerment as an adult, and my experience of teaching within the context of privilege—which, through the temporal and spatial, defines the moment and place of poverty, subordination and oppression—had established that "disadvantage" as "plain to see".

I began to re-realize that, in my initial thought-words of anger, I had been taking on the colonizing voice that produces the deficit, and that creates, validates, and establishes "the problem" from outside, from a place out there that can speak unmonitored by its own surveillance. I had been doing the same thing as that which I had surveyed in the courtyard. I was producing and reproducing the very conditions that produced the bully/ bullying in the first place, ensuring its reproduction through my own voyeuristic perspective and reproductive deficit language, albeit a silent language of thoughts.

I, too, had become a bully. I was complicit with a system or discourse and a well-entrenched paradigm of thinking that constructs "the problem," establishes the "truth" on "deficit," and *lays blame.*

I realized that my vantage point was at fault. That, in the contexts in which I had practiced my profession as a secondary school mathematics teacher in independent schools, the practices I was criticizing now would not have been possible—that teaching time was sacred, important, urgent, and that time was of the essence, a resource of which there was never enough. From *this* vantage point, I notice the inversion and the contradictions.

In a context of privilege, material resources are endless: paper, equipment, classroom space, computers, libraries, photocopy machines, photocopy assistants, new technology, availability of resource materials, curriculum materials, pedagogic assistance, all within a community of Pedagogic Knowledge Experts, but time was a precious and limited resource, urgent, sought after, coveted. Here, it was the opposite—confined space, discomfort, lack of privacy, lack of expertise, lack of pedagogic support, noise, dust, the smell of dust; too many bodies, huddled bodies, wriggling; no space to think, no space to prepare, but all the time in the world; in a sort of time-warp; in a mental, psychological, sociological, human landscape of *foreverness.*

These are the power principles that inform not only the political gaze from the perspective of the self, but also control the distributions of the spatial/temporal dichotomy and that define the political economy of context by assisting in the production of the poverty/ privilege hierarchy, and which define the roles of subjects in context.

These are the principles of the politics of context that either delimit or allow spaces of possibility, in accordance with the social division of labor of discourses (Bernstein, 2000) in the social domain in which the discourses of context are invested discourses that depend/suspend entirely on the stringently policed rules of relations of power.

This is how the principles of "progress" operate across the temporal and spatial, in the race towards "future advancement." And defined by its own rules, it ensures *very few* "winners," and *many* "losers"!

I hear the deficit voices again . . . bullying voices . . . some voices of educationalists, specialists, and well-known people in authority in South African Education; people in the "new arena" of post-liberation education; people I interviewed. "The problem lies with our teachers . . . they are underqualified, demotivated, lacking experience and expertise, and there is not enough of them. Our failures in mathematics can be directly attributed to the teachers; *they* are our problem."

I realize that, in my own way, I was feeding into this, re-creating this monster, re-establishing this deficit discourse. I realized that, in creating the teachers, principal, and their pedagogic practices in this "disadvantaged community" as *lacking*, as the "real problem", it was an escape, a way of not facing up to not understanding, not seeing the source of power and how it threads its way into the repressive web.

Yes, I had become the bully. And the bully in the courtyard was as much, if not more, my *victim* of constructed "disadvantage" and the pedagogy of pain and poverty that it produces as she was a bully in herself. The principal was a victim of it, too, and I had not even *begun* to imagine the strangulating and delimiting conditions that this discourse served to produce and in which he was constrained to operate. This was the "pedagogizing of difference," indeed, and a discourse in which I had participated. This was how the construction of disadvantage begot pedagogic and contextually produced disadvantage.

The principal came back into the room, looking a little harassed. A "sideshow" had interrupted our "performance," and had seemed to detract from "the conversation." But, in fact, it was a critical fragment of the whole, a necessary contribution to understanding the resolution of the narrative, in which our initial "polite" conversation preceding "the sideshow" had been the essential exposition. I, myself, had moved through several modes of *looking*, premised by various experiential podiums of perspective. Consequently, when I had been angry and critical, my vantage point had been the context of privilege in which I had gained much (although not all) of my teaching experience. When I had overcome my anger and realized my role in the co-constructed authorship of power, I had returned to my early youth and to remembering, remembering what it was like to be bullied and to feel the hand of violence and the voice of humiliation, and it was only *then* that I could begin to understand–feel with a *deeper listening*.

It had required a range of senses, as it had required a shift in perspective. I had moved from a "looking on" and the voyeuristic power instantiated in perspectives of "seeing," to a "listening to," where the eyes are quieted by the sights and sounds within darkened silence, and the sense of hearing is peaked, tuning *in to* silence.

This had been my route. Instead of trying to find the "root *of* the problem" and trying to "root *out* the problem," like a cancer from living tissue, instead, I was moving towards searching for "*the source.*" The source of the problem lay silently *behind* the construction of "the problem" itself, and threaded its way, like a tributary, to my very doorstep. I, too, was complicit, a collaborator of deficit discourse, a root of "the problem's" routedness. Now I became responsible as well, through acknowledging that responsibility.

The I–you dichotomy (see Buber, 1996, for "I–Thou" relationship) had been broken by the emergence of a new bond of responsibility...a *humbling togetherness*! I needed to *listen* collaboratively to that "source" in collectively finding a way together of "re-sourcing" towards non-impoverishment, other possibilities, and mutual healing.

On yet another level of perception, both ocular and audible, I realize that, in my criticism of the principal, I had been *not only* engaging in the reproduction of master narratives on poverty and deficit that lays the blame on the victim and not the discursive power base that establishes it, *but also* in the reproductive (re)creation of "Truth," or verisimilitude...and that the "truth" about the black teachers in South Africa was an act in the creation of a simulacrum. Where to now, in re-routing toward re-sourcing the discourse?

REFERENCES

Barta, J., & Bremner, M. (2009). Seeing with many eyes: Connections between anthropology and mathematics. In B. Greer, S. Mukhopadhyay, A. Powell, & S. Nelson-Barber (Eds.), *Culturally responsive mathematics education* (pp. 85–109). New York: Routledge.

Bernstein, B. (2000). *Pedagogy, symbolic control and identity: Theory, research and critique.* New York: Rowman & Littlefield.

Biesta, G. (2009). Good education: What it is and why we need it. Inaugural Lecture, The Stirling Institute of Education, 4th March 2009. University of Stirling, Scotland. [Online] Available: http://www.ioe.stir.ac.uk/documents/ GOODEDUCATION—WHATITISANDWHYWENEEDITInauguralLecture-ProfGertBiesta.pdf

Bishop, A. J. (1990). Western mathematics: The secret weapon of cultural imperialism. *Race and Class, 32*(2), 51 – 65.

Bishop, A. J. (2001). What values do you teach when you teach mathematics? In P. Gates (Ed.), *Issues in mathematics teaching* (pp. 93–104). London: Routledge Falmer.

Bourdieu, P., & Wacquant, J.D. (1992). *An invitation to reflexive sociology.* Chicago: University of Chicago Press.

Bowers, C. A. (2001). *Educating for eco-justice and community*. London: The University of Georgia Press.

Bowers, C. A., & Apffel-Marglin, F. (2005). *Rethinking Freire: Globalization and the environmental crisis*. Mahwah, New Jersey: Lawrence Erlbaum Associates.

Buber, M. (1996). *I and thou*. (Trans: W. Kaufmann). New York: Touchstone.

Connelly, F. M., & Clandinin, D. J. (1990). Stories of experience and narrative inquiry. *Educational Researcher, 14*(5), 2–14.

Connelly, F. M., & Clandinin, D. J. (1991). Narrative inquiry: Storied experience. In. E. Short (Ed.), *Forms of curriculum inquiry* (pp. 121–154). Albany, NY: State University of New York Press.

D'Ambrosio, U. (1997). Where does ethnomathematics stand nowadays? *For the Learning of Mathematics, 17*, 13–17.

D'Ambrosio, U. (2001). What is ethnomathematics, and how can it help children in schools? *Teaching Children Mathematics, 7*, 308–310.

D'Ambrosio, U. (2006). *Ethnomathematics: Link between tradition and modernity*. Rotterdam: Sense Publishers.

Deleuze, G., & Guattari, F. (1987). *A thousand plateaus: Capitalism and schizophrenia*. Minneapolis: University of Minnesota Press.

Donald, D. T. (2009). The curricular problem of indigenousness: Colonial frontier Logics, teacher resistances, and the acknowledgement of ethical space. In J. Nahachewsky & I. Johnston (Eds.), *Beyond 'presentism': Re-imagining the historical, personal, and social places of curriculum*. Rotterdam: Sense Publishers.

Dowling, P. (1998). *The sociology of mathematics education: Mathematical myths/pedagogic texts*. London: Falmer.

Dowling, P. (2001). Mathematics in late modernity: Beyond myths and fragmentation. In. B. Atweh, H. Forgasz, & B. Nebres (Eds.), *Sociocultural research on mathematics education: An international perspective* (pp. 19–36). Mahwah, NJ: Lawrence Erlbaum Associates.

Foucault, M. (1981). *The order of discourse* (Trans., R. Young). London: Routledge.

Freire, P. (1999). *Pedagogy of the oppressed*. New York: Continuum.

Gutstein, E. (2008). Building political relationships with students: An aspect of social justice pedagogy. In E. de Freitas & K Nolan (Eds.), *Opening the research text: Critical insights and in(ter)ventions into mathematics education* (pp. 189–204). New York: Springer.

Henderson, J. G., & Kesson, K. R. (2004). Preface. In J. G. Henderson & K. R. Kesson (Eds.), *Curriculum wisdom: Educational decisions in democratic societies*. New Jersey: Pearson/Merrill/Prentice Hall.

Jefferess, D. (2008). Global citizenship and the cultural politics of benevolence. *Critical Literacies: Theories and Practices, 2*(1). 27–36. [Online] Available: http://www.criticalliteracyjournal.org/

Lerman, S., & Tsatsaroni, A. (1998). Why children fail and what the field of mathematics education can do about it: The role of sociology. In P. Gates (Ed.), *Proceedings of the First International Mathematics Education and Society Conference* (pp. 26 – 33). Nottingham, UK: Centre for the Study of Mathematics Education, Nottingham University.

Pinar, W. F., Reynolds, W. M., Slattery, P., Taubman, P. M. (1996). *Understanding curriculum*. New York: Peter Lang.

Powell, A., & Frankenstein, M. (Eds.). (1997). *Ethnomathematics: Challenging eurocentrism in mathematics education*. Albany: State University of New York Press.

Sen, A. (1999). *Development as freedom*. New York: Anchor Books.

Skovsmose, O., & Valero, P. (2001). Breaking political neutrality: The critical engagement of mathematics education with democracy. In B. Atweh, H. Forgasz, & B. Nebres (Eds.), *Sociocultural research on mathematics education: An international perspective* (pp. 37–55). Mahwah, NJ: Lawrence Erlbaum Associates.

Swanson, D. M. (1998). *Bridging the boundaries? A study of mainstream mathematics, academic support and "disadvantaged learners" in an independent, secondary school in the western cape, (South Africa)*. Unpublished Master's dissertation, University of Cape Town, South Africa.

Swanson, D. M. (2000). Teaching mathematics in two independent school contexts: The construction of "good practice". *Educational Insights 6*(1) http://www.csci.educ.ubc.ca/publication/insights/online/v06n01/swanson.html

Swanson, D. M. (2001). "Disadvantage" and school mathematics: The politics of context. *The International Journal of Learning, 11*.

Swanson, D. M. (2004). *Voices in the Silence: Narratives of disadvantage, social context and school mathematics in post-apartheid South Africa*. Unpublished PhD dissertation, The University of British Columbia, Vancouver, Canada. [In press, Cambria Press].

Swanson, D. M. (2005). School mathematics, discourse and the politics of context. In A. Chronaki & I. Christiansen (Eds.), *Challenging perspectives on mathematics classroom communication* (pp. 261–294). Greenwich, CT: Information Age Publishing.

Swanson, D. M. (2006). *Power and poverty—Whose, where, and why?: School mathematics, context and the social construction of "disadvantage"*. In J. Novotná, H. Moraová, M. Kráatká, & N. Stehliková (Eds.), *Proceedings of the 30th conference of the International Group for the Psychology of Mathematics Education* (Vol. 5, pp. 217–224). Prague: PME.

Swanson, D. M. (2007a). Silent voices, silent bodies: Difference and disadvantage in schooling contexts. In D. Freedman & S. Springgay (Eds.), *Curriculum and the Cultural Body* (pp. 63–78). New York: Peter Lang. [Foreword by Madeleine Grumet.]

Swanson, D. M. (2007b). Shadows between us: An A/R/Tographic gaze at issues of ethics and activism. In S. Springgay, R. Irwin, C. Leggo, P. Gouzouasis, & K. Grauer (Eds.), *Being with a/r/tography* (pp. 179–185). Rotterdam: Sense Publishers.

Swanson, D. M. (2008). A Life for Aprile: Social justice as a search of/for the soul. A response to Eric Gutstein. In E. de Freitas & K. Nolan (Eds.), *Opening the Research Text: Critical Insights and In(ter)ventions into Mathematics Education* (pp. 213–219). New York: Springer.

Swanson, D. M. (2009, January). Roots / Routes (Part II), *Qualitative Inquiry, 15*(2) (Sage Publications)

Walker, D. F. (2003). *Fundamentals of curriculum: Passion and professionalism* (2nd ed.). Mahwah, NJ: Lawrence Erlbaum Associates.

Zevenbergen, R. (2003). Teachers' beliefs about teaching mathematics to students from socially-disadvantaged backgrounds: Implications for social justice. In

L. Burton (Ed.), *Which way social justice in mathematics education?* (pp. 133–151). Westport, CONN: Praeger.

FURTHER READING

Atweh, B., Forgasz, H., & Nebres, B. (Eds.). (2001). *Sociocultural research on mathematics education: An international perspective.* Mahwah, NJ: Lawrence Erlbaum Associates.

Bernstein, B. (2000). *Pedagogy, symbolic control and identity: Theory, research and critique.* New York: Rowman & Littlefield.

Dowling, P. (1998). *The sociology of mathematics education: Mathematical myths/pedagogic texts.* London: Falmer.

ABOUT THE CONTRIBUTORS

Tamara Bibby is a lecturer at the Institute of Education, London. She is interested in psychoanalytically informed and psychosocial explorations of the ways in which people come to make sense of themselves in their learning contexts. In relation to young people, this includes particularly their understandings of their learning relationships and processes, as well as group processes. This interest in identities also includes primary school teachers' views of themselves as professionals and their interactions with policy and their professional environments. Tamara's background in primary mathematics continues to influence the sites she chooses to focus on in her research and writing.

Tony Brown is Professor of Mathematics Education at the Manchester Metropolitan University. He has research interests in mathematics education and language, practitioner research in education, and teacher education. His books include: *Mathematics Education and Language: Interpreting Hermeneutics and Post-Structuralism*; *Action Research and Postmodernism* (with Liz Jones); *New Teacher Identity and Regulative Government* (with Olwen McNamara); *Regulative Discourses in Education: A Lacanian Perspective* (with Dennis Atkinson and Janice England); and an edited collection entitled *The Psychology of Mathematics Education: A Psychoanalytic Displacement*.

Nigel Calder is a senior lecturer in mathematics, science, and technology education at the University of Waikato, New Zealand. His research interests are digital technologies and their influence on the learning process in mathematics, mathematical thinking, generalization, and investigative

Unpacking Pedagogy: New Perspectives for Mathematics Classrooms, pages 265–269
265

approaches. He has taught in both primary and secondary school settings, is a former secondary mathematics adviser, and has been a contributor to many mathematics resource books. He is also interested in theorizing the research processes.

Tony Cotton is a mathematics teacher by trade, having taught mathematics in secondary schools for 10 years before moving to take on an advisory role in Multicultural Education. He has worked in publishing, with a responsibility for primary mathematics education, as well as in both the universities in Nottingham, UK. He has published books for pupils and their teachers, as well as a wide range of academic articles, and is a regular contributor to *Mathematics Teaching*, the professional journal for teachers of mathematics in the UK. He is currently Associate Dean and Head of Initial Teacher Education at Leeds Metropolitan University.

Brent Davis is Professor and Distinguished Research Chair in Mathematics Education at the University of Calgary, Canada. His research is developed around the educational relevance of developments in the cognitive and complexity sciences, and he teaches courses at the undergraduate and graduate levels in curriculum studies, mathematics education, and educational change. Brent has published books and articles in the areas of mathematics learning and teaching, curriculum theory, teacher education, epistemology, and action research. His most recent book is *Engaging Minds: Changing Teaching in Complex Times* (2nd ed., 2008; co-authored with Dennis Sumara and Rebecca Luce-Kapler).

Elizabeth de Freitas is an Associate Professor at Adelphi University, New York. Her research interests include mathematics education and cultural studies. She has published articles in *Educational Studies in Mathematics; Qualitative Inquiry; Race, Ethnicity and Education; Mathematics Teacher Education; The International Journal of Education and the Arts; Teaching Education; Language and Literacy; Gender and Education; The Journal of the Canadian Association for Curriculum Studies;* and *The Canadian Journal of Education.* She is also co-editor of the book *Opening the Research Text: Critical Insights and In(ter)ventions into Mathematics Education,* published by Springer Verlag in 2008.

Una Hanley is a senior lecturer in the Mathematics Centre at Manchester Metropolitan University, England. Her principal research interests are orientated around the development of critical methodology and around pedagogy—particularly, evolving teacher identities in a context of change. Her publications reflect this. Each of these strands marks a desire to find qualitative forms of understanding that offer alternative filters through which to examine aspects of teacher experience and the dilemmas immanent to it.

Kathleen Nolan is an Associate Professor in the Faculty of Education at the University of Regina, Canada, where she teaches undergraduate and graduate courses in mathematics curriculum and qualitative research. Her research interests include poststructural readings of mathematics education and narrative and self-study research methodologies. Her publications include articles in the *Journal of Mathematics Teacher Education* and the *Journal of Teaching and Learning*. She has pubished two books: *How Should I Know? Pre-service Teachers' Images of Knowing (by heart) in Mathematics and Science* (Sense Publishers) and *Opening the Research Text: Critical Insights and In(ter)ventions into Mathematics Education* (co-editor, Springer).

Ginny C. Powell is an Instructor of Mathematics at Georgia Perimeter College, Decatur, Georgia, USA, and a third-year PhD student in Mathematics Education at Georgia State University, Atlanta, Georgia, USA. Her research interests include remedial college-level mathematics education, democratic teaching, and the interplay of gender and ethnicity in mathematics learning. Her recent publications include "The View from Here: Opening up Postmodern Vistas" (book review of *Mathematics Education Within the Postmodern*) in *The Mathematics Educator*, and "Becoming Critical Mathematics Pedagogues: A Journey" (with David W. Stinson, Carla R. Bidwell, and Mary M. Thurman) in the *Proceedings of the Second Annual Meeting of the Georgia Association of Mathematics Teacher Educators*.

Moshe Renert is a doctoral candidate in Curriculum and Pedagogy at the University of British Columbia. His research focuses on relaxing orthodoxies of math education through increased awareness of the dialectical evolutionary processes that govern the field. As a math educator for the past 20 years, Moshe has worked with thousands of university, high school, and middle school students. He is currently the director of The Renert Centre, a chain of math schools in Western Canada. Previously, Moshe was an Assistant Professor of Computer Science.

Diana Stentoft has a PhD in mathematics education, and currently teaches and undertakes research in the Department of Education, Learning, and Philosophy at Aalborg University, Denmark. Her research is primarily concerned with investigating issues of complexity and instability in educational settings, with a focus on higher education in science and mathematics. She has a particular interest in issues of methodology and critical approaches in education and research.

David W. Stinson is Assistant Professor of Mathematics Education (and Women's Studies Institute Affiliated Faculty) at Georgia State University, Atlanta, Georgia, USA. His research examines how mathematics teachers, educators, and researchers incorporate the philosophical and theoretical underpinnings of critical postmodern theory into their education philosophies, pedagogical practices, and/or research methods; and how students, constructed outside the White, Christian, heterosexual male of bourgeois privilege, successfully accommodate, reconfigure, or resist (i.e., negotiate) the hegemonic discourses of schooling, and of society generally. His recent publications include articles in *The Mathematics Educator*, the *Review of Educational Research*, and the *American Educational Research Journal*.

Dalene Swanson holds Adjunct Professorships at The Universities of British Columbia and Alberta, Canada. Most recently, she was a postdoctoral scholar of the Social Sciences and Humanities Research Council of Canada. Dalene's research interests span mathematics education, curriculum studies, narrative methodologies, critical theory, indigeneity, and cultural studies. She has commitments to social, ecological, and global justice. Dalene holds degrees in mathematics and education from the University of Cape Town, and a PhD in Curriculum Studies from UBC. For her narrative-based dissertation addressing South African schools, Dalene received major Canadian, American, and international awards in Qualitative Research and Curriculum Studies. Web site: www.ualberta.ca/~dalene/index.html

Paola Valero is Associate Professor in Mathematics Education at the Department of Education, Learning, and Philosophy, Aalborg University, Denmark. She is leader of the Science and Mathematics Education Research Group and director of the Doctoral Program in Technology and Science: Education and Philosophy. Her research interests are mathematics, science, and engineering education, at all levels, in particular, innovation and change processes, curricular development, multiculturalism, and teacher education. She is the co-editor-in-chief of the journal *Nordic Studies in Mathematics Education*. She is co-editor of the book *Researching the Sociopolitical Dimensions of Mathematics Education: Issues of Power in Theory and Methodology* (2004).

Fiona Walls is a Senior Lecturer in mathematics education at James Cook University, North Queensland, Australia. She has worked as a Primary teacher in New Zealand, Australia, and Vanuatu, and as a pre-service teacher educator and professional development facilitator and adviser in mathematics education in New Zealand, Fiji, and Western Australia. Her research is focused on the cultural and historical construction of mathematics education, and the ways in which teachers, learners, and their communities experience mathematics education as a social act. Her book, *Mathematical Subjects: Children Talk about their Mathematics Lives* (Springer, 2009), examines the ways in which classrooms operate as social sites.

Margaret Walshaw is an Associate Professor at Massey University, New Zealand. She is co-director of the Centre of Excellence for Research in Mathematics Education (CERME) and coordinator of the Doctor of Education program at her university. She teaches postgraduate papers in research and in mathematics education. Margaret has published widely, both nationally and internationally. She is editor of the book *Mathematics Education within the Postmodern*, author of *Working with Foucault in Education*, co-author of *Are our Standards Slipping? Debates over Literacy and Numeracy Standards in New Zealand since 1945*, and co-author of *Effective Pedagogy in Mathematics/Pangarau*.

AUTHOR INDEX

Unpacking Pedagogy: New Perspectives for Mathematics Classrooms, pages 271–275
Copyright © 2010 by Information Age Publishing
271

SUBJECT INDEX